W9-BDB-461

Basic Computation

Finding Area and Perimeter

Loretta M. Taylor, Ed. D.
Mathematics Teacher
Hillsdale High School
San Mateo, California

Harold D. Taylor, Ed. D.
Head, Mathematics Department
Aragon High School
San Mateo, California

DALE
SEYMOUR
PUBLICATIONS
P.O. BOX 10888
PALO ALTO, CA 94303

Editors: Elaine C. Murphy, Susan McCalla
Production Coordinator: Ruth Cottrell
Cover designer: Michael Rogondino
Compositor: WB Associates
Printer: Malloy Lithographing

Copyright © 1981 by Dale Seymour Publications. All rights reserved. Printed in the United States of America. Published simultaneously in Canada.

Limited Reproduction Permission: The authors and publisher hereby grant permission to the teacher who purchases this book, or the teacher for whom the book is purchased, to reproduce up to 100 copies of any part of this book for use with his or her students. Any further duplication is prohibited.

ISBN 0-86651-006-0

Order Number DS01187

9 10 11 12 13 14 15-MA-95

DALE
SEYMOUR
PUBLICATIONS
P.O. BOX 10888
PALO ALTO, CA 94303

ABOUT THE PROGRAM

WHAT IS THE BASIC COMPUTATION LIBRARY?

The books in the BASIC COMPUTATION library together provide comprehensive practice in all the essential computational skills. There are practice books and a test book. The practice books consist of carefully sequenced drill worksheets organized in groups of five. The test book contains daily quizzes (160 quizzes in all), semester tests, and year-end tests written in standardized-test formats.

If you find this book effective, you may want to use others in the series. Build your own library to suit your own needs.

BOOK 1	WORKING WITH WHOLE NUMBERS
BOOK 2	UNDERSTANDING FRACTIONS
BOOK 3	WORKING WITH FRACTIONS
BOOK 4	WORKING WITH DECIMALS
BOOK 5	WORKING WITH PERCENTS
BOOK 6	UNDERSTANDING MEASUREMENT
BOOK 7	FINDING AREA AND PERIMETER
BOOK 8	WORKING WITH CIRCLES AND VOLUME
BOOK 9	APPLYING COMPUTATIONAL SKILLS
TEST BOOK	BASIC COMPUTATION QUIZZES AND TESTS

WHO CAN USE THE BASIC COMPUTATION LIBRARY?

Classroom teachers, substitute teachers, tutors, parents, and persons wishing to study on their own can use these materials. Although written specifically for the general math classroom, books in the BASIC COMPUTATION library can be used with any program requiring carefully sequenced computational practice. The material is appropriate for use with any person, young or old, who has not yet certified computational proficiency. It is especially suitable for middle school, junior high school, and high school students who need to master the essential computational skills necessary for mathematical literacy.

WHAT IS IN THIS BOOK?

This book is a practice book. In addition to these teacher notes, it contains student worksheets, example problems, and a record form.

Worksheets

The worksheets are designed to give even the slowest student a chance to master the essential computational skills. Most worksheets come in five equivalent forms allowing for pretesting, practice, and posttesting on any one skill. Each set of worksheets provides practice in only one or two specific skills and the work progresses in very small steps from one set to the next. Instructions are clear and simple, with handwritten samples of the exercises completed. Ample practice is provided on each page, giving students the opportunity to strengthen their skills. Answers to each problem are included in the back of the book.

Example Problems

Fully-worked examples show how to work each type of exercise. Examples are keyed to the worksheet pages. The example solutions are written in a straightforward manner and are easily understood.

Record Form

A record form is provided to help in recording progress and assessing instructional needs.

Answers

Answers to each problem are included in the back of the book.

HOW CAN THE BASIC COMPUTATION LIBRARY BE USED?

The materials in the BASIC COMPUTATION library can serve as the major skeleton of a skills program or as supplements to any other computational skills program. The large number of worksheets gives a wide variety from which to choose and allows flexibility in structuring a program to meet individual needs. The following suggestions are offered to show how the BASIC COMPUTATION library may be adapted to a particular situation.

Minimal Competency Practice

In various fields and schools, standardized tests are used for entrance, passage from one level to another, and certification of competency or proficiency prior to graduation. The materials in the BASIC COMPUTATION library are particularly well-suited to preparing for any of the various mathematics competency tests, including the mathematics portion of the General Educational Development test (GED) used to certify high school equivalency.

Together, the books in the BASIC COMPUTATION library give practice in all the essential computational skills measured on competency tests. The semester tests and year-end tests from the test book are written in standardized-test formats. These tests can be used as sample minimal competency tests. The worksheets can be used to brush up on skills measured by the competency tests.

Skill Maintenance

Since most worksheets come in five equivalent forms, the computation work can be organized into weekly units as suggested by the following schedule. Day one is for pretesting and introducing a skill. The next three days are for drill and practice followed by a unit test on the fifth day.

AUTHORS' SUGGESTED TEACHING SCHEDULE

	Day 1	Day 2	Day 3	Day 4	Day 5
Week 1	pages 1 and 2 pages 11 and 12	pages 3 and 4 pages 13 and 14	pages 5 and 6 pages 15 and 16	pages 7 and 8 pages 17 and 18	pages 9 and 10 pages 19 and 20
Week 2	pages 21 and 22 pages 31 and 32	pages 23 and 24 pages 33 and 34	pages 25 and 26 pages 35 and 36	pages 27 and 28 pages 37 and 38	pages 29 and 30 pages 39 and 40
Week 3	pages 41 and 42 pages 51 and 52	pages 43 and 44 pages 53 and 54	pages 45 and 46 pages 55 and 56	pages 47 and 48 pages 57 and 58	pages 49 and 50 pages 59 and 60
Week 4	pages 61 and 62 pages 71 and 72	pages 63 and 64 pages 73 and 74	pages 65 and 66 pages 75 and 76	pages 67 and 68 pages 77 and 78	pages 69 and 70 pages 79 and 80

The daily quizzes from BASIC COMPUTATION QUIZZES AND TESTS can be used on the drill and practice days for maintenance of previously-learned skills or diagnosis of skill deficiencies.

A five-day schedule can begin on any day of the week. The authors' ideal schedule begins on Thursday, with reteaching on Friday. Monday and Tuesday are for touch-up teaching and individualized instruction. Wednesday is test day.

Supplementary Drill

There are more than 18,000 problems in the BASIC COMPUTATION library. When students need more practice with a given skill, use the appropriate worksheets from the library. They are suitable for classwork or homework practice following the teaching of a specific skill. With five equivalent pages for most worksheets, adequate practice is provided for each essential skill.

HOW ARE MATERIALS PREPARED?

The books are designed so the pages can be easily removed and reproduced by Thermofax, Xerox, or a similar process. For example, a ditto master can be made on a Thermofax for use on a spirit duplicator. Permanent transparencies can be made by processing special transparencies through a Thermofax or Xerox.

Any system will run more smoothly if work is stored in folders. Record forms can be attached to the folders so that either students or teachers can keep records of individual progress. Materials stored in this way are readily available for conferences.

EXAMPLE PROBLEMS

AREA AND PERIMETER OF RECTANGLES

EXAMPLE   Find the length, width, perimeter, and area of the rectangle.

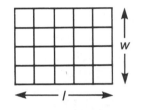

Solution: length = 5, width = 4
Perimeter = $2l + 2w = 18$
Area = $l \times w = 20$

MORE AREA AND PERIMETER OF RECTANGLES

EXAMPLE   Find the length, width, perimeter, and area of the rectangle.

Solution: length = 4, width = 1
Perimeter = $2l + 2w = 10$
Area = $l \times w = 4$

DRAWING RECTANGLES

EXAMPLE 1 Use graph paper to draw 3 rectangles, each with a different shape, having area 30. Give the dimensions and perimeter of each.

Solution: $A = 30$, so $l \times w = 30$. Three possibilities are:
a. $l = 15, w = 2, P = 34$ **b.** $l = 10, w = 3, P = 26$
c. $l = 6, w = 5, P = 22$

EXAMPLE 2 Use graph paper to draw 3 rectangles, each with a different shape, having perimeter 26. Give the dimensions and area of each.

Solution: $P = 26$, so $2l + 2w = 26$, or $l + w = 13$. Three possibilities are:
a. $l = 10, w = 3, A = 30$ **b.** $l = 8, w = 5, A = 40$
c. $l = 7, w = 6, A = 42$

MORE DRAWING RECTANGLES

EXAMPLE   Use graph paper to draw a rectangle having area 20 and perimeter 24.

Solution: To have area 20, a rectangle could have dimensions 1 by 20, or 2 by 10, or 4 by 5. The only one of these having perimeter 24 is the 2 by 10 rectangle. So the rectangle has length = 10 and width = 2.

ENGLISH UNITS FOR RECTANGLES

EXAMPLE   Find the area and perimeter of a rectangle having length 4 ft and width 1 yd.

Solution: Length and width must be in the same units. Change width to 1 yd to width to 3 ft. Then, $P = 2l + 2w = 14$ ft and $A = l \times w = 12$ ft².

METRIC UNITS FOR RECTANGLES

EXAMPLE   Find the area and perimeter of a rectangle having length 17 cm and width 15 mm.

Solution: Length and width must be in the same units. Change the length from 17 cm to 170 mm. Then, $P = 2l + 2w = 370$ mm and $A = l \times w = 2550$ mm².

USING ENGLISH UNITS FOR RECTANGLES

EXAMPLE   Find the width and area of a rectangle having length 24 in. and perimeter 82 in.

Solution: Since $P = 82$, $2l + 2w = 82$, $l + w = 41$, and $w = 41 - 24 = 17$ in. Then, $A = l \times w = 408$ in.².

USING METRIC UNITS FOR RECTANGLES

EXAMPLE   Find the length and perimeter of a rectangle having width 43 cm and area 528.9 cm².

Solution: $A = l \times w$, so $l = A \div w = 528.9 \div 43 = 12.3$ cm. Then, $P = 2l + 2w = 110.6$ cm.

AREA OF RIGHT TRIANGLES

EXAMPLE Find the base, altitude, and area of the right triangle.

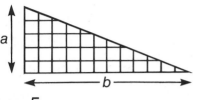

Solution: base = 12, altitude = 5

$$\text{Area} = \frac{1}{2} \times b \times a = \frac{1}{2} \times 12 \times 5$$
$$= 30$$

MORE AREA OF RIGHT TRIANGLES

EXAMPLE Use graph paper to draw 3 right triangles, each with a different shape, having area 45.

Solution: Since $A = \frac{1}{2}(b \times a)$, $2A = b \times a$. Then $b \times a = 2 \times 45$ = 90. Three possibilities are:

a. $b = 1, a = 90$ **b.** $b = 2, a = 45$ **c.** $b = 3, a = 30$

ENGLISH UNITS FOR RIGHT TRIANGLES

EXAMPLE Find the area and perimeter of the right triangle having sides as follows: $a = 3$ ft, $b = 4$ ft, and $c = 5$ ft.

Solution: $A = \frac{1}{2}(b \times a) = \frac{1}{2}(4 \times 3) = 6$ ft
$P = a + b + c = 12$ ft

METRIC UNITS FOR RIGHT TRIANGLES

EXAMPLE Find the area and perimeter of the right triangle having sides as follows: $a = 10$ m, $b = 24$ m, and $c = 26$ m.

Solution: $A = \frac{1}{2}(b \times a) = \frac{1}{2}(24 \times 10) = 120$ m²
$P = a + b + c = 60$ m

AREA OF TRIANGLES

EXAMPLE Find the base, altitude, and area of the triangle.

Solution: base = 16, altitude = 8

$$\text{Area} = \frac{1}{2}(b \times a) = \frac{1}{2}(16 \times 8) = 64$$

DRAWING TRIANGLES

EXAMPLE 1 Use graph paper to draw triangles, each with a different shape, having base 6 and altitude 3. Give the area for each.

Solution: Each triangle will have area $\frac{1}{2}(6 \times 3)$ or 9.

EXAMPLE 2 Use graph paper to draw triangles, each with a different shape, having area 8. Give the base and altitude for each.

Solution: Since $A = \frac{1}{2}(b \times a)$, $2A = b \times a$. So, $b \times a = 2 \times 8 = 16$. One possibility is $b = 4$, $a = 4$.

AREA OF TRIANGLES AND PARALLELOGRAMS

EXAMPLE 1 Find the area and perimeter of the triangle having sides as follows: $a = 10$, $b = 21$, $c = 17$, and altitude $h = 8$.

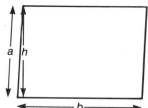

Solution: $A = \frac{1}{2}(b \times h) = \frac{1}{2}(21 \times 8) = 84$
$P = a + b + c = 48$

EXAMPLE 2 Find the area and perimeter of the parallelogram having sides as follows: $a = 61$, $b = 81$, and altitude $h = 60$.

Solution: $A = b \times h = 81 \times 60 = 4860$
$P = 2a + 2b = 2(61) + 2(81) = 284$

AREA OF TRAPEZOIDS

EXAMPLE Find the area and perimeter of the trapezoid having sides as follows: $a = 34$, $b_1 = 42$, $c = 50$, $b_2 = 90$, and altitude $h = 30$.

Solution: $A = \frac{1}{2}h(b_1 + b_2) = \frac{1}{2} \times 30 \times (42 + 90) = 1980$
$P = a + b_1 + c + b_2 = 216$

STUDENT RECORD SHEET

Worksheets Completed

Page Number

1	3	5	7
2	4	6	8
11	13	15	17
12	14	16	18
21	23	25	27
22	24	26	28
31	33	35	37
32	34	36	38
41	43	45	47
42	44	46	48
51	53	55	57
52	54	56	58
61	63	65	67
62	64	66	68
71	73	75	77
72	74	76	78

9
10
19
20
29
30
39
40
49
50
59
60
69
70
79
80

Daily Quiz Grades

No.	Score

Check List
Skill Mastered

Date

☐ area & perimeter of rectangles

☐ drawing rectangles

☐ English units for rectangles

☐ metric units for rectangles

☐ using metric units for rectangles

☐ area of right triangles

☐ English units for right triangles

☐ metric units for right triangles

☐ area of triangles

☐ drawing triangles

☐ area of triangles & parallelograms

☐ area of trapezoids

Notes

_____

_____

_____

_____

_____

_____

_____

_____

Worksheet page _____   Name _____

Date _____

Area and perimeter of rectangles

Name _____

Date _____

Find the length, width, perimeter, and area of each rectangle.

1. length = _10_
   width = _10_
   perimeter = _40_
   area = _100_

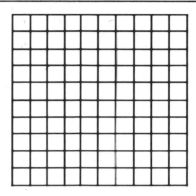

2. length = _____
   width = _____
   perimeter = _____
   area = _____

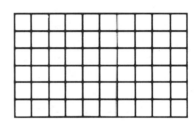

3. length = _____
   width = _____
   perimeter = _____
   area = _____

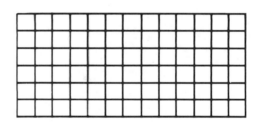

4. length = _____
   width = _____
   perimeter = _____
   area = _____

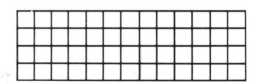

5. length = _____
   width = _____
   perimeter = _____
   area = _____

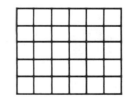

Copyright © 1981 by Dale Seymour Publications.

More area and perimeter of rectangles

Name _____

Date _____

Find the length, width, perimeter, and area of each rectangle.

1. length = _9_
   width = _3_
   perimeter = _24_
   area = _27_

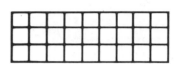

2. length = _____
   width = _____
   perimeter = _____
   area = _____

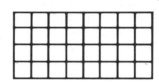

3. length = _____
   width = _____
   perimeter = _____
   area = _____

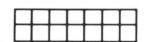

4. length = _____
   width = _____
   perimeter = _____
   area = _____

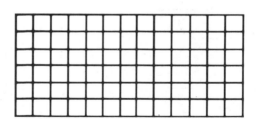

5. length = _____
   width = _____
   perimeter = _____
   area = _____

Copyright © 1981 by Dale Seymour Publications.

Area and perimeter of rectangles

Name _____

Date _____

Find the length, width, perimeter, and area of each rectangle.

1. length = _12_
width = _15_
perimeter = _54_
area = _180_

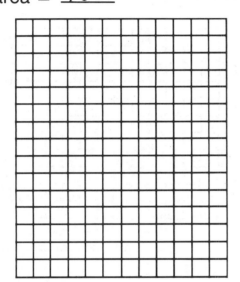

2. length = _____
width = _____
perimeter = _____
area = _____

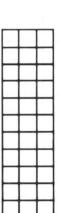

3. length = _____, width = _____, perimeter = _____, area = _____

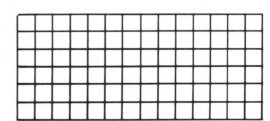

4. length = _____
width = _____
perimeter = _____
area = _____

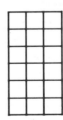

5. length = _____
width = _____
perimeter = _____
area = _____

Copyright © 1981 by Dale Seymour Publications.

Name _____

Date _____

Find the length, width, perimeter, and area of each rectangle.

1. length = ___6___
width = ___11___
perimeter = ___34___
area = ___66___

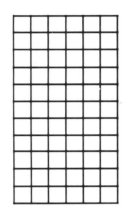

2. length = _____
width = _____
perimeter = _____
area = _____

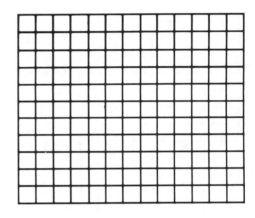

3. length = _____
width = _____
perimeter = _____
area = _____

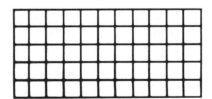

4. length = _____
width = _____
perimeter = _____
area = _____

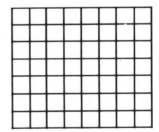

5. length = _____, width = _____, perimeter = _____, area = _____

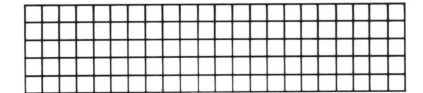

Copyright © 1981 by Dale Seymour Publications.

Area and perimeter of rectangles

Name _____

Date _____

Find the length, width, perimeter, and area of each rectangle.

1. length = _12_
width = _6_
perimeter = _36_
area = _72_

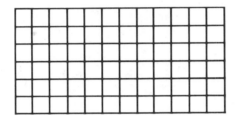

2. length = _____
width = _____
perimeter = _____
area = _____

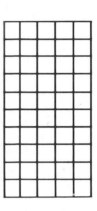

3. length = _____, width = _____, perimeter = _____, area = _____

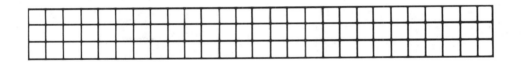

4. length = _____
width = _____
perimeter = _____
area = _____

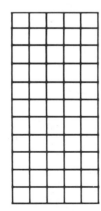

5. length = _____
width = _____
perimeter = _____
area = _____

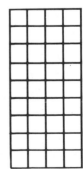

Copyright © 1981 by Dale Seymour Publications.

More area and perimeter of rectangles

Name _____

Date _____

Find the length, width, perimeter, and area of each rectangle.

1. length = __6__
width = __15__
perimeter = __42__
area = __90__

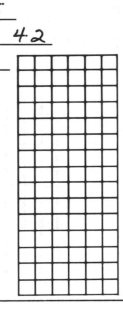

2. length = _____
width = _____
perimeter = _____
area = _____

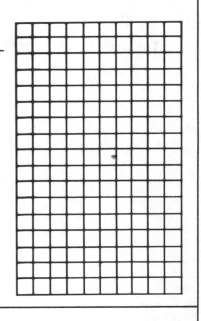

3. length = _____
width = _____
perimeter = _____
area = _____

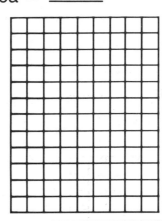

4. length = _____
width = _____
perimeter = _____
area = _____

5. length = _____, width = _____, perimeter = _____, area = _____

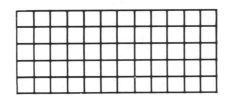

Copyright © 1981 by Dale Seymour Publications.

Area and perimeter of rectangles

Name _____

Date _____

Find the length, width, perimeter, and area of each rectangle.

1. length = _11_
width = _16_
perimeter = _54_
area = _176_

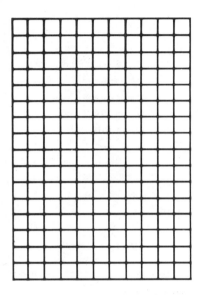

2. length = _____
width = _____
perimeter = _____
area = _____

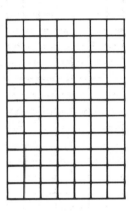

3. length = _____
width = _____
perimeter = _____
area = _____

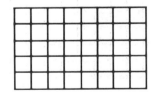

4. length = _____
width = _____
perimeter = _____
area = _____

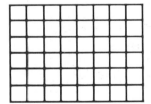

5. length = _____, width = _____, perimeter = _____, area = _____

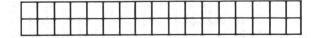

Copyright © 1981 by Dale Seymour Publications.

More area and perimeter of rectangles

Name _____

Date _____

Find the length, width, perimeter, and area of each rectangle.

1. length = _7_
width = _14_
perimeter = _42_
area = _98_

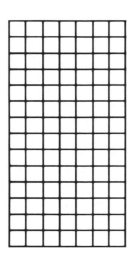

2. length = _____
width = _____
perimeter = _____
area = _____

3. length = _____
width = _____
perimeter = _____
area = _____

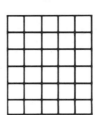

4. length = _____
width = _____
perimeter = _____
area = _____

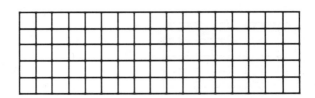

5. length = _____
width = _____
perimeter = _____
area = _____

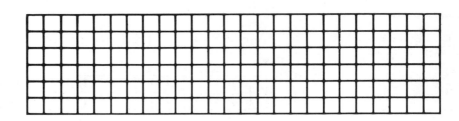

Copyright © 1981 by Dale Seymour Publications.

Area and perimeter of rectangles

Name _____

Date _____

Find the length, width, perimeter, and area of each rectangle.

1. length = _11_
width = _10_
perimeter = _42_
area = _110_

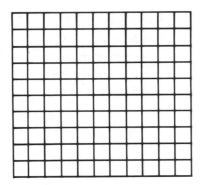

2. length = _____
width = _____
perimeter = _____
area = _____

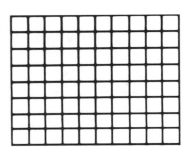

3. length = _____
width = _____
perimeter = _____
area = _____

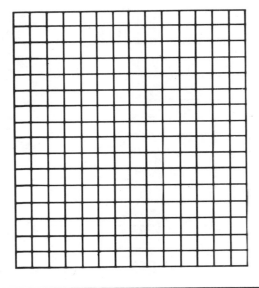

4. length = _____
width = _____
perimeter = _____
area = _____

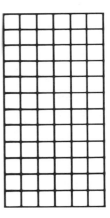

5. length = _____, width = _____, perimeter = _____, area = _____

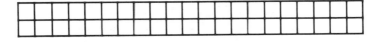

Copyright © 1981 by Dale Seymour Publications.

More area and perimeter of rectangles

Name _____

Date _____

Find the length, width, perimeter, and area of each rectangle.

1. length = _12_
width = _12_
perimeter = _48_
area = _144_

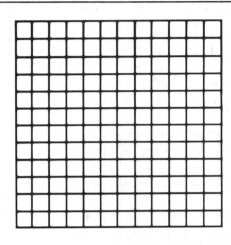

2. length = _____
width = _____
perimeter = _____
area = _____

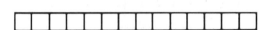

3. length = _____
width = _____
perimeter = _____
area = _____

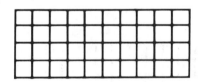

4. length = _____
width = _____
perimeter = _____
area = _____

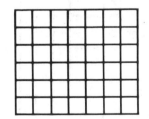

5. length = _____
width = _____
perimeter = _____
area = _____

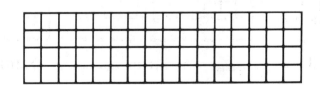

Copyright © 1981 by Dale Seymour Publications.

Drawing rectangles

Name _____

Date _____

Use graph paper to draw 3 rectangles with different shapes
and having the given area.
Write the dimensions and perimeter of each rectangle in the chart.

	area	first rectangle			second rectangle			third rectangle		
		l	*w*	P	*l*	*w*	P	*l*	*w*	P
1.	12 squares	6	2	16	4	3	14	12	1	26
2.	20 squares									
3.	24 squares									
4.	18 squares									
5.	30 squares									

Use graph paper to draw 3 rectangles with different shapes
and having the given perimeter.
Write the dimensions and area of each rectangle in the chart.

	perimeter	first rectangle			second rectangle			third rectangle		
		l	*w*	A	*l*	*w*	A	*l*	*w*	A
6.	30 units	8	7	56	9	6	54	10	5	50
7.	38 units									
8.	28 units									
9.	22 units									
10.	26 units									

Copyright © 1981 by Dale Seymour Publications.

More drawing rectangles

Name _____

Date _____

One graph paper, draw a rectangle having the given area and perimeter.
Write the dimensions of each rectangle on the chart.

	perimeter	area	length	width
1.	22 units	28 squares	7	4
2.	26 units	36 squares		
3.	26 units	40 squares		
4.	30 units	56 squares		
5.	28 units	45 squares		

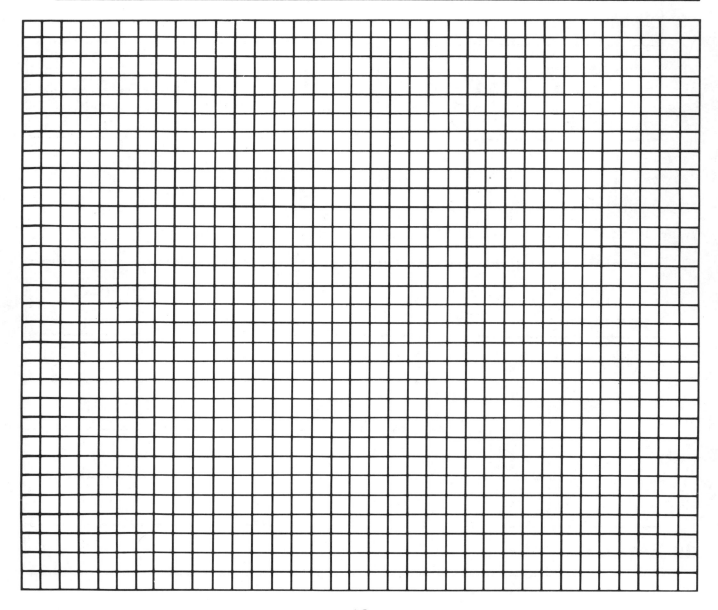

Copyright © 1981 by Dale Seymour Publications.

Drawing rectangles

Use graph paper to draw 3 rectangles with different shapes
and having the given area.
Write the dimensions and perimeter of each rectangle in the chart.

	area	first rectangle			second rectangle			third rectangle		
		l	*w*	*P*	*l*	*w*	*P*	*l*	*w*	*P*
1.	60 squares	10	6	32	15	4	38	12	5	34
2.	100 squares									
3.	32 squares									
4.	42 squares									
5.	50 squares									

Use graph paper to draw 3 rectangles with different shapes
and having the given perimeter.
Write the dimensions and area of each rectangle in the chart.

	perimeter	first rectangle			second rectangle			third rectangle		
		l	*w*	*A*	*l*	*w*	*A*	*l*	*w*	*A*
6.	40 units	19	1	19	18	2	36	17	3	51
7.	56 units									
8.	14 units									
9.	32 units									
10.	46 units									

Copyright © 1981 by Dale Seymour Publications.

13

More drawing rectangles

On graph paper, draw a rectangle having the given area and perimeter.
Write the dimensions of each rectangle on the chart.

	perimeter	area	length	width
1.	30 units	36 squares	12	3
2.	28 units	24 squares		
3.	36 units	45 squares		
4.	22 units	30 squares		
5.	18 units	20 squares		

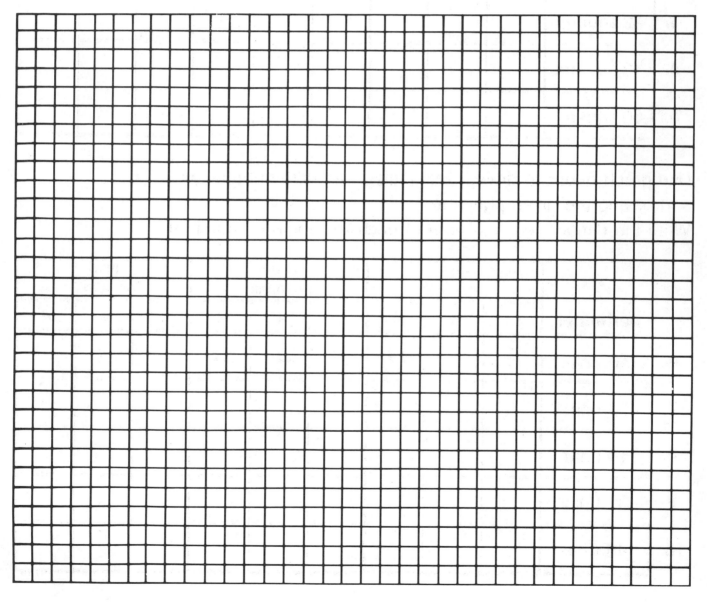

Copyright © 1981 by Dale Seymour Publications.

Drawing rectangles

Use graph paper to draw 3 rectangles with different shapes
and having the given area.
Write the dimensions and perimeter of each rectangle in the chart.

area	first rectangle			second rectangle			third rectangle		
	l	*w*	*P*	*l*	*w*	*P*	*l*	*w*	*P*
1. 28 squares	28	1	58	14	2	32	7	4	22
2. 32 squares									
3. 30 squares									
4. 42 squares									
5. 75 squares									

Use graph paper to draw 3 rectangles with different shapes
and having the given perimeter.
Write the dimensions and area of each rectangle in the chart.

perimeter	first rectangle			second rectangle			third rectangle		
	l	*w*	*A*	*l*	*w*	*A*	*l*	*w*	*A*
6. 18 units	5	4	20	8	1	8	7	2	14
7. 20 units									
8. 34 units									
9. 48 units									
10. 22 units									

Copyright © 1981 by Dale Seymour Publications.

15

Name _____

Date _____

On graph paper, draw a rectangle having the given area and perimeter.
Write the dimensions of each rectangle on the chart.

	perimeter	area	length	width
1.	20 units	25 squares	5	5
2.	46 units	90 squares		
3.	30 units	44 squares		
4.	22 units	24 squares		
5.	34 units	60 squares		

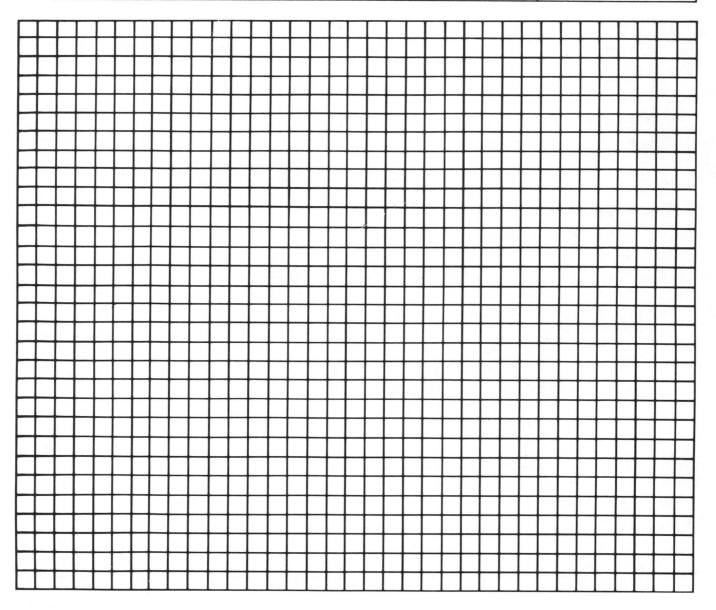

Copyright © 1981 by Dale Seymour Publications.

16

Drawing rectangles

Name _____

Date _____

Use graph paper to draw 3 rectangles with different shapes
and having the given area.
Write the dimensions and perimeter of each rectangle in the chart.

	area	first rectangle			second rectangle			third rectangle		
		l	*w*	P	*l*	*w*	P	*l*	*w*	P
1.	36 squares	36	1	74	18	2	40	9	4	26
2.	30 squares									
3.	40 squares									
4.	72 squares									
5.	54 squares									

Use graph paper to draw 3 rectangles with different shapes
and having the given perimeter.
Write the dimensions and area of each rectangle in the chart.

	perimeter	first rectangle			second rectangle			third rectangle		
		l	*w*	A	*l*	*w*	A	*l*	*w*	A
6.	16 units	7	1	7	6	2	12	5	3	15
7.	24 units									
8.	36 units									
9.	28 units									
10.	32 units									

Copyright © 1981 by Dale Seymour Publications.

More drawing rectangles

Name _____

Date _____

On graph paper, draw a rectangle having the given area and perimeter.
Write the dimensions of each rectangle on the chart.

	perimeter	area	length	width
1.	46 units	60 squares	20	3
2.	36 units	81 squares		
3.	24 units	36 squares		
4.	48 units	80 squares		
5.	38 units	84 squares		

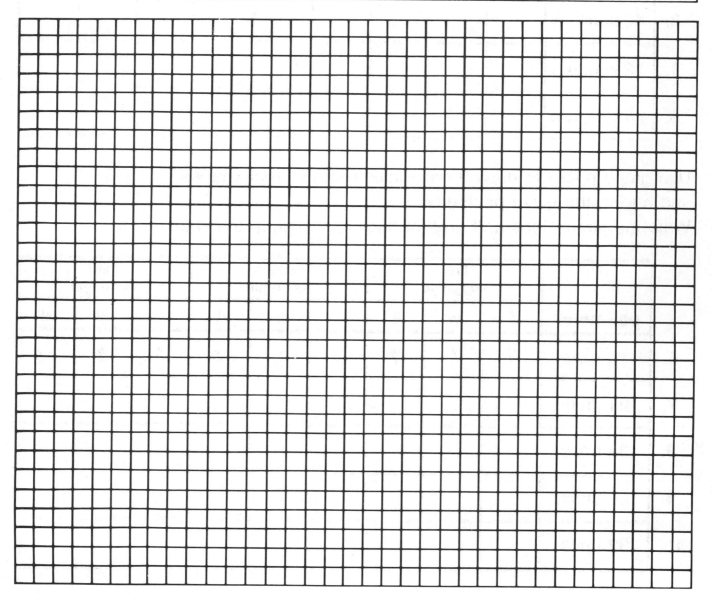

Copyright © 1981 by Dale Seymour Publications.

18

Drawing rectangles

Name _____

Date _____

Use graph paper to draw 3 rectangles with different shapes
and having the given area.
Write the dimensions and perimeter of each rectangle in the chart.

		first rectangle		second rectangle			third rectangle		
area		l	w	P	l	w	P	l	w	P
1.	20 squares	20	1	42	10	2	24	5	4	18
2.	45 squares									
3.	32 squares									
4.	27 squares									
5.	18 squares									

Use graph paper to draw 3 rectangles with different shapes
and having the given perimeter.
Write the dimensions and area of each rectangle in the chart.

		first rectangle		second rectangle			third rectangle		
perimeter		l	w	A	l	w	A	l	w	A
6.	12 units	5	1	5	4	2	8	3	3	9
7.	18 units									
8.	22 units									
9.	16 units									
10.	14 units									

Copyright © 1981 by Dale Seymour Publications.

More drawing rectangles

Name _____

Date _____

On graph paper, draw a rectangle having the given area and perimeter.
Write the dimensions of each rectangle on the chart.

	perimeter	area	length	width
1.	26 units	40 squares	8	5
2.	20 units	24 squares		
3.	32 units	64 squares		
4.	30 units	56 squares		
5.	20 units	16 squares		

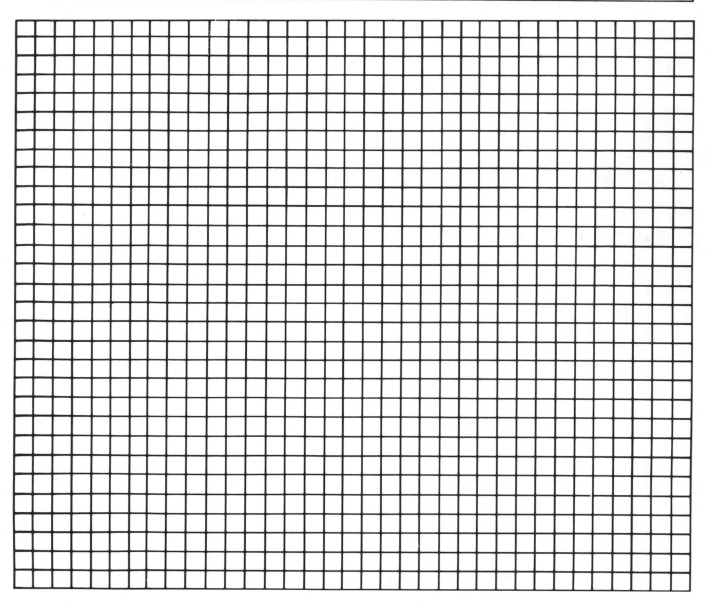

Copyright © 1981 by Dale Seymour Publications.

English units for rectangles

Name _____

Date _____

The following are formulas for
area and perimeter of a rectangle.

$$A = l \times w \qquad P = 2l + 2w$$

Use the formulas to find the areas
and perimeters of rectangles with the
given dimensions.
*means you must change units.

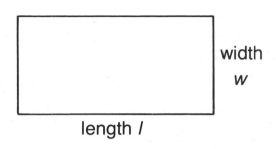

width
w

length l

	length	width	perimeter	area
1.	12 ft	24 ft	72 ft	288 ft²
2.	35 yd	10 yd		
3.	35 in.	49 in.		
4.*	27 yd	24 ft		
5.*	12 ft	8 ft 6 in.		

Copyright © 1981 by Dale Seymour Publications.

Metric units for rectangles

Name _____

Date _____

The following are formulas for
area and perimeter of a rectangle.

$$A = l \times w \qquad P = 2l + 2w$$

Use the formulas to find the areas
and perimeters of rectangles with the
given dimensions.

*means you must change units.

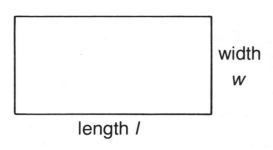

width
w

length l

	length	width	perimeter	area
1.	1.9 m	0.26 m	4.32 m	0.494 m²
2.	1.7 m	0.32 m		
3.*	2.3 cm	32 m		
4.*	2.5 km	410 m		
5.*	16 mm	0.27 cm		

Copyright © 1981 by Dale Seymour Publications.

English units for rectangles

Name _____

Date _____

The following are formulas for
area and perimeter of a rectangle.

$$A = l \times w \qquad P = 2l + 2w$$

Use the formulas to find the areas
and perimeters of rectangles with the
given dimensions.
*means you must change units.

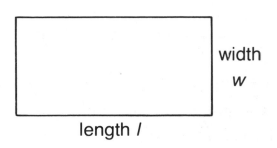

width
w

length l

	length	width	perimeter	area
1.	35 in.	18 in.	106 in.	630 in.²
2.	22 ft	17 ft		
3.*	8 ft	96 in.		
4.*	9 ft	8 ft 4 in.		
5.*	8 ft	5 ft 3 in.		

Copyright © 1981 by Dale Seymour Publications.

23

Metric units for rectangles

The following are formulas for
area and perimeter of a rectangle.

$$A = l \times w \qquad P = 2l + 2w$$

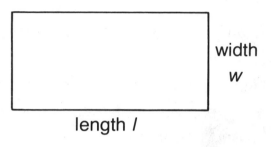

width
w

length l

Use the formulas to find the areas
and perimeters of rectangles with the
given dimensions.
*means you must change units.

	length	width	perimeter	area
1.	9.3 dm	7.2 dm	33 dm	66.92 dm²
2.	87 cm	92 cm		
3.*	3.6 m	120 cm		
4.*	12.2 dkm	2.5 m		
5.*	11.4 m	1.25 cm		

right © 1981 by Dale Seymour Publications.

24

English units for rectangles

Name _____

Date _____

The following are formulas for
area and perimeter of a rectangle.

$A = l \times w$ $P = 2l + 2w$

Use the formulas to find the areas
and perimeters of rectangles with the
given dimensions.
*means you must change units.

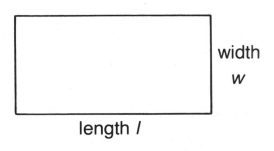

length *l*

width
w

	length	width	perimeter	area
1.	12 ft	8 ft	40 ft	96 ft²
2.	15 ft	9 ft		
3.*	9 ft 3 in.	12 ft		
4.*	6 yd	7 yd 8 in.		
5.*	17 in.	2 ft		

Copyright © 1981 by Dale Seymour Publications.

Metric units for rectangles

Name _____

Date _____

The following are formulas for
area and perimeter of a rectangle.

$$A = l \times w \qquad P = 2l + 2w$$

Use the formulas to find the areas
and perimeters of rectangles with the
given dimensions.
*means you must change units.

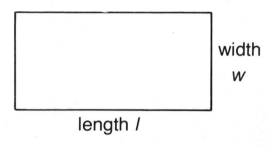

length *l*

	length	width	perimeter	area
1.	1.7 dm	5.3 dm	14 dm	9.01 dm^2
2.*	4.3 m	12 dm		
3.*	8.6 cm	1.2 mm		
4.*	15 dkm	23 m		
5.*	14 m	0.023 mm		

Copyright © 1981 by Dale Seymour Publications.

English units for rectangles

The following are formulas for
area and perimeter of a rectangle.

$$A = l \times w \qquad P = 2l + 2w$$

Use the formulas to find the areas
and perimeters of rectangles with the
given dimensions.
*means you must change units.

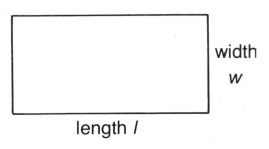

length *l*

width
w

	length	width	perimeter	area
1.	14 in.	17 in.	62 in.	238 in.²
2.*	12 ft	5 ft 6 in.		
3.*	35 ft	14 yd		
4.*	13 ft	5 yd		
5.	15 ft 2 in.	18 ft		

Copyright © 1981 by Dale Seymour Publications.

Metric units for rectangles

The following are formulas for
area and perimeter of a rectangle.

$$A = l \times w \qquad P = 2l + 2w$$

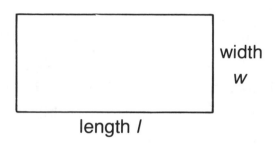

width
w

length l

Use the formulas to find the areas
and perimeters of rectangles with the
given dimensions.
*means you must change units.

	length	width	perimeter	area
1.	9.2 dm	8.7 dm	35.8 dm	80.04 dm²
2.*	4.3 m	28 dm		
3.*	2.9 cm	85 mm		
4.*	23 cm	8.1 m		
5.*	6.5 m	23 cm		

Copyright © 1981 by Dale Seymour Publications.

English units for rectangles

Name _____

Date _____

The following are formulas for
area and perimeter of a rectangle.

$$A = l \times w \qquad P = 2l + 2w$$

Use the formulas to find the areas
and perimeters of rectangles with the
given dimensions.
*means you must change units.

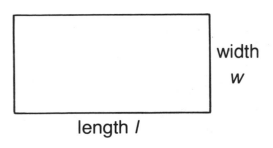

width
w

length l

	length	width	perimeter	area
1.	43 in.	25 in.	136 in.	1075 in.^{2}
2.	96 ft	73 ft		
3.*	19 yd	27 ft		
4.*	28 ft	6 ft 6 in.		
5.*	16 ft	58 ft 3 in.		

Copyright © 1981 by Dale Seymour Publications.

Metric units for rectangles

The following are formulas for
area and perimeter of a rectangle.

$$A = l \times w \qquad P = 2l + 2w$$

Use the formulas to find the areas
and perimeters of rectangles with the
given dimensions.
*means you must change units.

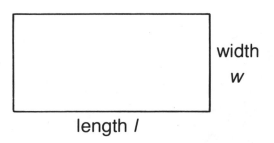

width
w

length l

	length	width	perimeter	area
1.	1.6 cm	0.28 cm	3.76 cm	0.448 cm²
2.	5.9 mm	8.6 mm		
3.*	13.2 m	125 cm		
4.*	6.5 dm	63 cm		
5.*	82 m	0.43 cm		

Copyright © 1981 by Dale Seymour Publications.

Using English units for rectangles

Name _____

Date _____

Complete the following table. Use $A = l \times w$ and $P = 2l + 2w$.

	length	width	perimeter	area
1.	30 ft	8 ft	76 ft	240 ft²
2.	32 yd	12 yd	_____	_____
3.	72 in.	16 in.	_____	_____
4.	27 in.	14 in.	_____	_____
5.	62 ft	15 ft	_____	_____
6.	$4\frac{1}{4}$ mi	$6\frac{2}{5}$ mi	_____	_____
7.	5 ft 3 in.	5 ft 4 in.	_____	_____
8.	19 in.	_____	_____	494 in.²
9.	29 ft	_____	_____	522 ft²
10.	_____	14 ft	238 ft	_____
11.	_____	36 yd	328 yd	_____

Copyright © 1981 by Dale Seymour Publications.

31

Using metric units for rectangles

Name _____

Date _____

Complete the following table. Use $A = l \times w$ and $P = 2l + 2w$.

	length	width	perimeter	area
1.	12.3 cm	9.2 cm	_43 cm_	_113.16 cm²_
2.	7.32 m	4.01 m	_____	_____
3.	_____	3.2 m	_____	18.24 m²
4.	69 mm	_____	_____	278.07 mm²
5.	3.5 cm	_____	12.2 cm	_____
6.	8.7 m	_____	36 m	_____
7.	46 cm	_____	_____	190.44 cm²

Copyright © 1981 by Dale Seymour Publications.

Using English units for rectangles

Name _____

Date _____

Complete the following table. Use $A = l \times w$ and $P = 2l + 2w$.

	length	width	perimeter	area
1.	15 ft	5 ft	_40 ft_	_75 ft²_
2.	16 in.	12 in.	_____	_____
3.	40 yd	20 yd	_____	_____
4.	23 ft	15 ft	_____	_____
5.	17 in.	32 in.	_____	_____
6.	3 ft 4 in.	2 ft 3 in.	_____	_____
7.	6 yd 1 ft	7 yd 18 in.	_____	_____
8.	19 ft	_____	_____	133 ft²
9.	26 in.	_____	_____	312 in.²
10.	_____	24 yd	118 yd	_____
11.	57 ft	_____	178 ft	_____

Copyright © 1981 by Dale Seymour Publications.

33

Using metric units for rectangles

Name _____

Date _____

Complete the following table. Use $A = l \times w$ and $P = 2l + 2w$.

	length	width	perimeter	area
1.	83 cm	72 cm	_310 cm_	_5976 cm²_
2.	57 mm	36 mm	_____	_____
3.	10.3 m	8.2 m	_____	_____
4.	5.7 dm	3.4 dm	_____	_____
5.	4.8 m	_____	_____	12.48 m²
6.	5.11 cm	_____	_____	18.3449 cm²
7.	_____	3.9 m	18.6 m	_____

Copyright © 1981 by Dale Seymour Publications.

34

Using English units for rectangles   Name _____

                                                  Date _____

Complete the following table. Use $A = l \times w$ and $P = 2l + 2w$.

	length	width	perimeter	area
1.	28 ft	13 ft	_82 ft_	_364 ft²_
2.	58 ft	16 ft	_____	_____
3.	72 in.	34 in.	_____	_____
4.	57 yd	26 yd	_____	_____
5.	10 ft 6 in.	7 ft 4 in.	_____	_____
6.	6 ft 8 in.	4 ft 9 in.	_____	_____
7.	9 yd 9 in.	11 yd 12 in.	_____	_____
8.	43 in.	_____	_____	1204 in.²
9.	67 in.	_____	_____	2881 in.²
10.	_____	14 in.	82 in.	_____
11.	15 ft	_____	40 ft	_____

Copyright © 1981 by Dale Seymour Publications.

Using metric units for rectangles

Name _____

Date _____

Complete the following table. Use $A = l \times w$ and $P = 2l + 2w$.

	length	width	perimeter	area
1.	15 m	5 m	_40 m_	_75 m²_
2.	24 cm	12 cm	_____	_____
3.	37 cm	2.9 cm	_____	_____
4.	_____	51 cm	_____	3723 cm²
5.	_____	63 mm	_____	6048 mm²
6.	77 cm	_____	222 cm	_____
7.	38 mm	_____	128 mm	_____

Copyright © 1981 by Dale Seymour Publications.

Using English units for rectangles

Name _____

Date _____

Complete the following table. Use $A = l \times w$ and $P = 2l + 2w$.

	length	width	perimeter	area
1.	48 ft	12 ft	*120 ft*	*576 ft²*
2.	37 in.	19 in.	_____	_____
3.	39 in.	29 in.	_____	_____
4.	76 ft	32 ft	_____	_____
5.	125 yd	14 yd	_____	_____
6.	5 ft 4 in.	2 ft 3 in.	_____	_____
7.	3 ft 4 in.	4 ft 3 in.	_____	_____
8.	7 yd 2 ft	3 yd 1 ft	_____	_____
9.	29 in.	_____	_____	377 in.²
10.	86 ft	_____	_____	2064 ft²
11.	_____	33 in.	_____	561 in.²

Copyright © 1981 by Dale Seymour Publications.

Using metric units for rectangles

Name _____

Date _____

Complete the following table. Use $A = l \times w$ and $P = 2l + 2w$.

	length	width	perimeter	area
1.	58 mm	26 mm	*168mm*	*1508mm²*
2.	18.2 cm	7.5 cm	_____	_____
3.	0.927 m	1.3 m	_____	_____
4.	8.6 mm	5.32 mm	_____	_____
5.	92 cm	_____	_____	3174 cm²
6.	69 m	_____	544 m	_____
7.	62 mm	_____	178 mm	_____

Copyright © 1981 by Dale Seymour Publications.

Area of right triangles

Name _____

Date _____

Find the base, altitude, and area of each right triangle.

1. base = *11*
altitude = *4*
area = *22*

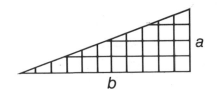

2. base = _____
altitude = _____
area = _____

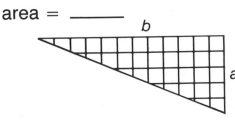

3. base = _____
altitude = _____
area = _____

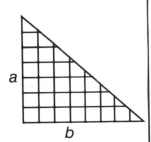

4. base = _____
altitude = _____
area = _____

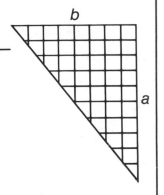

5. base = _____
altitude = _____
area = _____

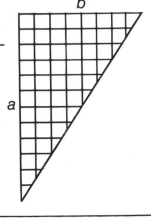

6. base = _____
altitude = _____
area = _____

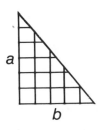

7. base = _____
altitude = _____
area = _____

8. base = _____
altitude = _____
area = _____

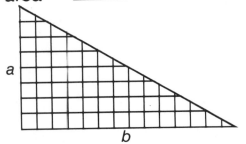

Copyright © 1981 by Dale Seymour Publications.

More area of right triangles

Name _____

Date _____

Find the base, altitude, and area of each right triangle.

1. base = _16_
    altitude = _8_
    area = _64_

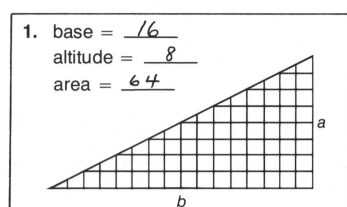

2. base = _____
    altitude = _____
    area = _____

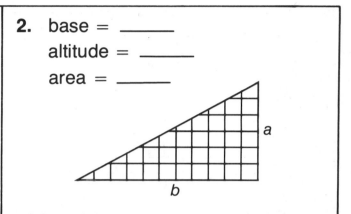

Use graph paper to draw 3 right triangles, each with a different shape and having an area of 36 square units. Give the base and altitude for each.

3. base = _____
    altitude = _____

4. base = _____
    altitude = _____

5. base = _____
    altitude = _____

Use graph paper to draw 3 right triangles, each with a different shape and having an area of 24 square units. Give the base and altitude for each.

6. base = _____
    altitude = _____

7. base = _____
    altitude = _____

8. base = _____
    altitude = _____

Copyright © 1981 by Dale Seymour Publications.

Area of right triangles

Find the base, altitude, and area of each right triangle.

1. base = _11_

altitude = _6_

area = _33_

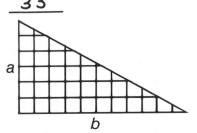
a
b

2. base = _____

altitude = _____

area = _____

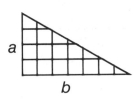
a
b

3. base = _____

altitude = _____

area = _____

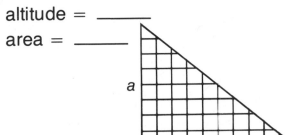
a
b

4. base = _____

altitude = _____

area = _____

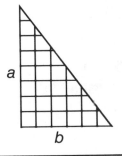
a
b

5. base = _____

altitude = _____

area = _____

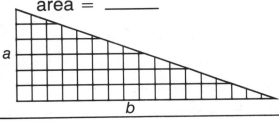
a
b

6. base = _____

altitude = _____

area = _____

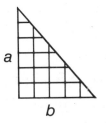
a
b

7. base = _____

altitude = _____

area = _____

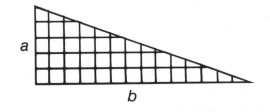
a
b

8. base = _____

altitude = _____

area = _____

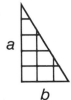
a
b

Copyright © 1981 by Dale Seymour Publications.

43

More area of right triangles   Name _____

                                             Date _____

Find the base, altitude, and area of each right triangle.

1. base = _5_

  altitude = _4_

  area = _10_

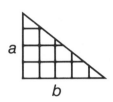

a

b

2. base = _____

  altitude = _____

  area = _____

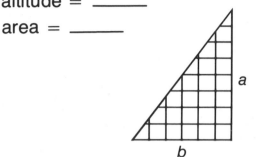

a

b

Use graph paper to draw 3 right triangles, each with a different shape and having an area of 30 square units. Give the base and altitude for each.

3. base = _____

  altitude = _____

4. base = _____

  altitude = _____

5. base = _____

  altitude = _____

Use graph paper to draw 3 right triangles, each with a different shape and having an area of 40 square units. Give the base and altitude for each.

6. base = _____

  altitude = _____

7. base = _____

  altitude = _____

8. base = _____

  altitude = _____

Copyright © 1981 by Dale Seymour Publications.

Area of right triangles

Name _____

Date _____

Find the base, altitude, and area of each right triangle.

1. base = _13_
altitude = _6_
area = _39_

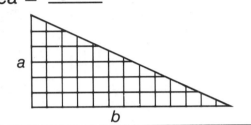
a *b*

2. base = _____
altitude = _____
area = _____

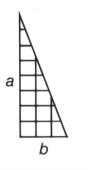
a *b*

3. base = _____
altitude = _____
area = _____

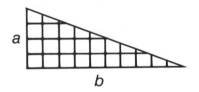
a *b*

4. base = _____
altitude = _____
area = _____

a *b*

5. base = _____
altitude = _____
area = _____

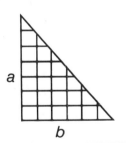
a *b*

6. base = _____
altitude = _____
area = _____

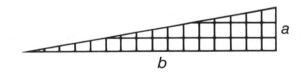
a *b*

7. base = _____
altitude = _____
area = _____

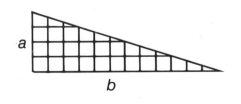
a *b*

8. base = _____
altitude = _____
area = _____

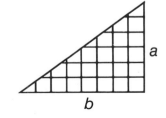
a *b*

Copyright © 1981 by Dale Seymour Publications.

More area of right triangles

Name _____

Date _____

Find the base, altitude, and area of each right triangle.

1. base = __9__
altitude = __7__
area = __$31\frac{1}{2}$__

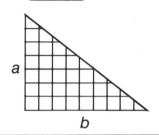

2. base = _____
altitude = _____
area = _____

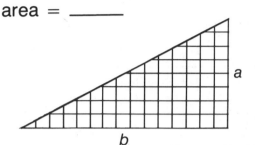

Use graph paper to draw 3 right triangles, each with a different shape and having an area of 12 square units. Give the base and altitude for each.

3. base = _____
altitude = _____

4. base = _____
altitude = _____

5. base = _____
altitude = _____

Use graph paper to draw 3 right triangles, each with a different shape and having an area of 8 square units. Give the base and altitude for each.

6. base = _____
altitude = _____

7. base = _____
altitude = _____

8. base = _____
altitude = _____

Copyright © 1981 by Dale Seymour Publications.

Area of right triangles

Name _____

Date _____

Find the base, altitude, and area of each right triangle.

1. base = _13_
altitude = _6_
area = _39_

2. base = _____
altitude = _____
area = _____

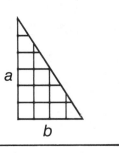

3. base = _____
altitude = _____
area = _____

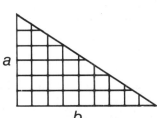

4. base = _____
altitude = _____
area = _____

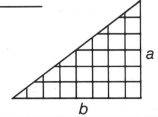

5. base = _____
altitude = _____
area = _____

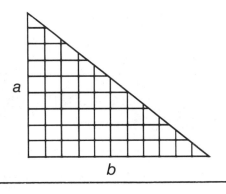

6. base = _____
altitude = _____
area = _____

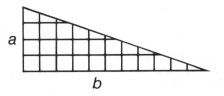

7. base = _____
altitude = _____
area = _____

8. base = _____
altitude = _____
area = _____

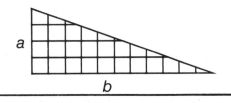

Copyright © 1981 by Dale Seymour Publications.

More area of right triangles

Name _____

Date _____

Find the base, altitude, and area of each right triangle.

1. base = _4_
   altitude = _4_
   area = _8_

2. base = _____
   altitude = _____
   area = _____

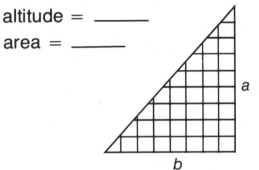

Use graph paper to draw 3 right triangles, each with a different shape and having an area of 10 square units. Give the base and altitude for each.

3. base = _____
   altitude = _____

4. base = _____
   altitude = _____

5. base = _____
   altitude = _____

Use graph paper to draw 3 right triangles, each with a different shape and having an area of 20 square units. Give the base and altitude for each.

6. base = _____
   altitude = _____

7. base = _____
   altitude = _____

8. base = _____
   altitude = _____

Copyright © 1981 by Dale Seymour Publications.

Area of right triangles

Name _____

Date _____

Find the base, altitude, and area of each right triangle.

1. base = _11_

altitude = _9_

area = _49½_

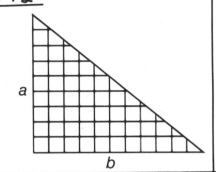

2. base = _____

altitude = _____

area = _____

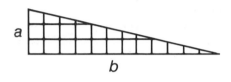

3. base = _____

altitude = _____

area = _____

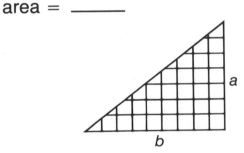

4. base = _____

altitude = _____

area = _____

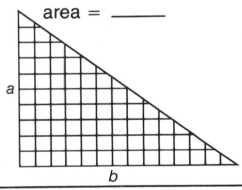

5. base = _____

altitude = _____

area = _____

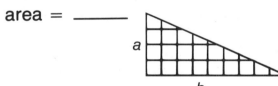

6. base = _____

altitude = _____

area = _____

7. base = _____

altitude = _____

area = _____

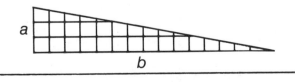

8. base = _____

altitude = _____

area = _____

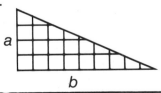

Copyright © 1981 by Dale Seymour Publications.

Name _____

Date _____

Find the base, altitude, and area of each right triangle.

1. base = _14_
altitude = _2_
area = _14_

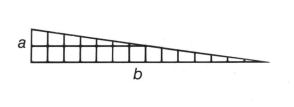

2. base = _____
altitude = _____
area = _____

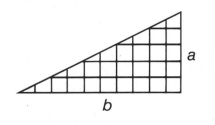

Use graph paper to draw 3 right triangles, each with a different shape and having an area of 18 square units. Give the base and altitude for each.

3. base = _____
altitude = _____

4. base = _____
altitude = _____

5. base = _____
altitude = _____

Use graph paper to draw 3 right triangles, each with a different shape and having an area of 16 square units. Give the base and altitude for each.

6. base = _____
altitude = _____

7. base = _____
altitude = _____

8. base = _____
altitude = _____

Copyright © 1981 by Dale Seymour Publications.

English units for right triangles

Name _____

Date _____

Formulas for the area and perimeter of a right triangle are as follows:

area = $\frac{1}{2}$ altitude × base   perimeter = sum of sides
$A = \frac{1}{2} ab$ $P = a + b + c$

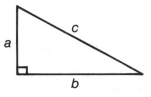

Find the area and perimeter of right triangles having the given dimensions.

1. $a = 20$ in.
   $b = 21$ in.
   $c = 29$ in.
   area = _210 in.²_
   perimeter = _70 in._

2. $a = 28$ yd
   $b = 45$ yd
   $c = 53$ yd
   area = _____
   perimeter = _____

3. $a = 48$ ft
   $b = 55$ ft
   $c = 73$ ft
   area = _____
   perimeter = _____

4. $a = 60$ in.
   $b = 91$ in.
   $c = 109$ in.
   area = _____
   perimeter = _____

5. $a = 40$ yd
   $b = 42$ yd
   $c = 58$ yd
   area = _____
   perimeter = _____

Copyright © 1981 by Dale Seymour Publications.

Formulas for the area and perimeter of a right triangle are as follows:

area = $\frac{1}{2}$ altitude × base   perimeter = sum of sides

$A = \frac{1}{2} ab$ $P = a + b + c$

Find the area and perimeter of right triangles having the
given dimensions.

*means you must change units.

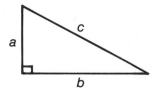

1. a = 33 km
b = 56 km
c = 65 km
area = $\underline{924\,km^2}$
perimeter = $\underline{154\,km}$

2. a = 11 m
b = 60 m
c = 61 m
area = _____
perimeter = _____

3. a = 84 cm
b = 135 cm
c = 159 cm
area = _____
perimeter = _____

4.* a = 39 cm
b = 0.8 m
c = 8.9 dm
area = _____
perimeter = _____

5.* a = 0.13 m
b = 84 cm
c = 8.5 dm
area = _____
perimeter = _____

Copyright © 1981 by Dale Seymour Publications.

Name _____

Date _____

Formulas for the area and perimeter of a right triangle are as follows:

area = $\frac{1}{2}$ altitude × base   perimeter = sum of sides
$A = \frac{1}{2}ab$ $P = a + b + c$

Find the area and perimeter of right triangles having the given dimensions.

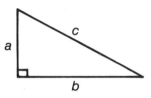

1. a = 20 ft
   b = 99 ft
   c = 101 ft
   area = _990 ft²_
   perimeter = _220 ft_

2. a = 40 in.
   b = 42 in.
   c = 58 in.
   area = _____
   perimeter = _____

3. a = 15 yd
   b = 20 yd
   c = 25 yd
   area = _____
   perimeter = _____

4. a = 36 in.
   b = 77 in.
   c = 85 in.
   area = _____
   perimeter = _____

5. a = 33 ft
   b = 56 ft
   c = 65 ft
   area = _____
   perimeter = _____

Copyright © 1981 by Dale Seymour Publications.

Metric units for right triangles

Name _____

Date _____

Formulas for the area and perimeter of a right triangle are as follows:

area = $\frac{1}{2}$ altitude × base   perimeter = sum of sides

$A = \frac{1}{2}ab$ $P = a + b + c$

Find the area and perimeter of right triangles having the
given dimensions.
*means you must change units.

1. a = 3.3 cm
   b = 4.4 cm
   c = 5.5 cm
   area = __7.26 cm^2__
   perimeter = __13.2 cm__

2. a = 4.0 m
   b = 9.6 m
   c = 10.4 m
   area = _____
   perimeter = _____

3. a = 5.6 dm
   b = 10.5 dm
   c = 11.9 dm
   area = _____
   perimeter = _____

4.* a = 6.3 m
    b = 2160 cm
    c = 225 dm
    area = _____
    perimeter = _____

5.* a = 54 cm
    b = 2.4 m
    c = 2460 mm
    area = _____
    perimeter = _____

Copyright © 1981 by Dale Seymour Publications.

English units for right triangles

Name _____

Date _____

Formulas for the area and perimeter of a right triangle are as follows:

area = $\frac{1}{2}$ altitude × base   perimeter = sum of sides
$A = \frac{1}{2} ab$ $P = a + b + c$

Find the area and perimeter of right triangles having the given dimensions.

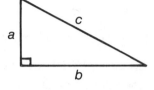

*means you must change units.

---

1. a = 10 ft
b = 24 ft
c = 26 ft
area = *120 ft²*
perimeter = *60 ft*

---

2. a = 27 ft
b = 36 ft
c = 45 ft
area = _____
perimeter = _____

---

3. a = 60 in.
b = 144 in.
c = 156 in.
area = _____
perimeter = _____

---

4. a = 10 yd
b = $10\frac{1}{2}$ yd
c = $14\frac{1}{2}$ yd
area = _____
perimeter = _____

5.* a = 81 ft
b = 120 yd
c = 123 yd
area = _____
perimeter = _____

Copyright © 1981 by Dale Seymour Publications.

Metric units for right triangles

Name _____

Date _____

Formulas for the area and perimeter of a right triangle are as follows:

area = ½ altitude × base   perimeter = sum of sides
$A = \frac{1}{2} ab$ $P = a + b + c$

Find the area and perimeter of right triangles having the
given dimensions.

*means you must change units.

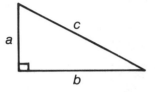

---

1. a = 3.3 cm
b = 4.4 cm
c = 5.5 cm
area = __7.26 cm²__
perimeter = __13.2 cm__

---

2. a = 7.2 mm
b = 13.5 mm
c = 15.3 mm
area = _____
perimeter = _____

---

3. a = 0.56 km
b = 1.92 km
c = 2 km
area = _____
perimeter = _____

---

4. a = 8.8 dkm
b = 16.5 dkm
c = 18.7 dkm
area = _____
perimeter = _____

5.* a = 0.45 km
b = 600 m
c = 750 m
area = _____
perimeter = _____

Copyright © 1981 by Dale Seymour Publications.

English units for right triangles

Formulas for the area and perimeter of a right triangle are as follows:

area = $\frac{1}{2}$ altitude × base

$A = \frac{1}{2} ab$

perimeter = sum of sides

$P = a + b + c$

Find the area and perimeter of right triangles having the given dimensions.

*means you must change units.

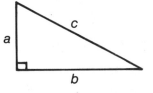

1. a = 18 in.
   b = 24 in.
   c = 30 in.
   area = __216 in.²__
   perimeter = __72 in.__

2. a = 24 ft
   b = 32 ft
   c = 40 ft
   area = _____
   perimeter = _____

3. a = 72 yd
   b = 320 yd
   c = 328 yd
   area = _____
   perimeter = _____

4. a = 40 in.
   b = 96 in.
   c = 104 in.
   area = _____
   perimeter = _____

5.* a = 14 ft
    b = 576 in.
    c = 50 ft
    area = _____
    perimeter = _____

Copyright © 1981 by Dale Seymour Publications.

Metric units for right triangles

Name _____

Date _____

Formulas for the area and perimeter of a right triangle are as follows:

area = $\frac{1}{2}$ altitude × base   perimeter = sum of sides
    $A = \frac{1}{2} ab$ $P = a + b + c$

Find the area and perimeter of right triangles having the given dimensions.

*means you must change units.

1. a = 5.4 m
   b = 7.2 m
   c = 9 m
   area = _19.44 m²_
   perimeter = _21.6 m_

2. a = 0.7 km
   b = 1.68 km
   c = 1.82 km
   area = _____
   perimeter = _____

3. a = 3.6 km
   b = 4.8 km
   c = 6 km
   area = _____
   perimeter = _____

4. a = 12 dm
   b = 12.6 dm
   c = 17.4 dm
   area = _____
   perimeter = _____

5.* a = 3.9 m
   b = 520 cm
   c = 6500 mm
   area = _____
   perimeter = _____

Copyright © 1981 by Dale Seymour Publications.

58

English units for right triangles   Name _____

                                          Date _____

Formulas for the area and perimeter of a right triangle are as follows:

area = $\frac{1}{2}$ altitude × base   perimeter = sum of sides
A = $\frac{1}{2}$ ab   P = a + b + c

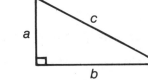

Find the area and perimeter of right triangles having the
given dimensions.
*means you must change units.

---

1. a = 16 ft
   b = 30 ft
   c = 34 ft
   area = ___240 ft²___
   perimeter = ___80 ft___

---

2. a = 18 in.
   b = 80 in.
   c = 82 in.
   area = _____
   perimeter = _____

---

3. a = 36 yd
   b = 77 yd
   c = 85 yd
   area = _____
   perimeter = _____

---

4. a = 65 ft
   b = 72 ft
   c = 97 ft
   area = _____
   perimeter = _____

5.* a = 13 in.
   b = 7 ft
   c = 85 in.
   area = _____
   perimeter = _____

Copyright © 1981 by Dale Seymour Publications.

Metric units for right triangles

Name _____

Date _____

Formulas for the area and perimeter of a right triangle are as follows:

area = $\frac{1}{2}$ altitude × base
$A = \frac{1}{2} ab$

perimeter = sum of sides
$P = a + b + c$

Find the area and perimeter of right triangles having the given dimensions.

*means you must change units.

1. $a = 21$ cm
$b = 28$ cm
$c = 35$ cm
area = _294 cm^2_
perimeter = _84 cm_

2. $a = 14$ mm
$b = 48$ mm
$c = 50$ mm
area = _____
perimeter = _____

3. $a = 3.3$ cm
$b = 5.6$ cm
$c = 6.5$ cm
area = _____
perimeter = _____

4. $a = 0.11$ dm
$b = 0.60$ dm
$c = 0.61$ dm
area = _____
perimeter = _____

5.* $a = 3.9$ m
$b = 0.008$ km
$c = 890$ cm
area = _____
perimeter = _____

Copyright © 1981 by Dale Seymour Publications.

Area of triangles

Name _____

Date _____

Find the base, altitude, and area of each triangle.

1. base = _9_

altitude = _4_

area = _18_

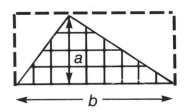

2. base = _____

altitude = _____

area = _____

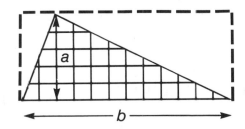

3. base = _____

altitude = _____

area = _____

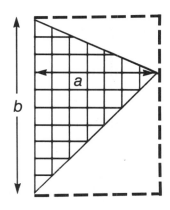

4. base = _____

altitude = _____

area = _____

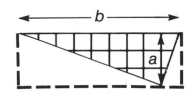

5. base = _____

altitude = _____

area = _____

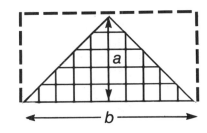

Copyright © 1981 by Dale Seymour Publications.

61

Drawing triangles

Use graph paper to draw triangles, each with a different shape, having the given base and altitude. Make one a right triangle. Give the area for each.

1. base = 6 units altitude = 5 units area = _15 sq units_	2. base = 9 units altitude = 4 units area = _____	3. base = 12 units altitude = 3 units area = _____

Use graph paper to draw triangles, each with a different shape, having the given area. Give the base and altitude for each.

4. area = 14 sq units base = _____ altitude = _____	5. area = 16 sq units base = _____ altitude = _____	6. area = 10 sq units base = _____ altitude = _____

Copyright © 1981 by Dale Seymour Publications.

Area of triangles

Find the base, altitude, and area of each triangle.

1. base = _11_
altitude = _5_
area = _$27\frac{1}{2}$_

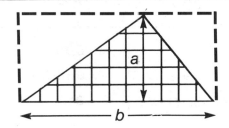

2. base = _____
altitude = _____
area = _____

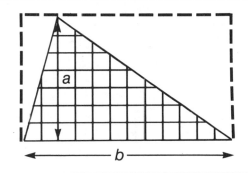

3. base = _____
altitude = _____
area = _____

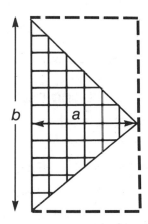

4. base = _____
altitude = _____
area = _____

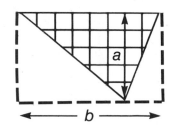

5. base = _____
altitude = _____
area = _____

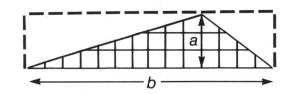

Copyright © 1981 by Dale Seymour Publications.

Drawing triangles

Name _____

Date _____

Use graph paper to draw triangles, each with a different shape, having the given base and altitude. Make one a right triangle. Give the area for each.

1. base = 7 units	**2.** base = 10 units	**3.** base = 12 units
altitude = 4 units	altitude = 9 units	altitude = 12 units
area = *14 sq units*	area = _____	area = _____

Use graph paper to draw triangles, each with a different shape, having the given area. Give the base and altitude for each.

4. area = 20 sq units	**5.** area = 18 sq units	**6.** area = 12 sq units
base = _____	base = _____	base = _____
altitude = _____	altitude = _____	altitude = _____

Copyright © 1981 by Dale Seymour Publications.

Name _____

Date _____

Find the base, altitude, and area of each triangle.

1. base = _13_

altitude = _5_

area = _32½_

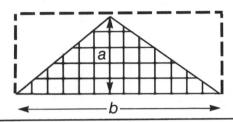

2. base = _____

altitude = _____

area = _____

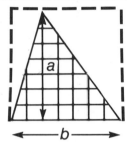

3. base = _____

altitude = _____

area = _____

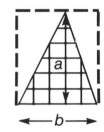

4. base = _____

altitude = _____

area = _____

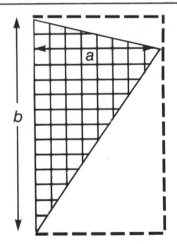

5. base = _____

altitude = _____

area = _____

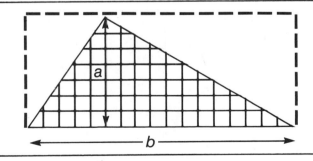

Copyright © 1981 by Dale Seymour Publications.

Drawing triangles

Use graph paper to draw triangles, each with a different shape, having the given base and altitude. Make one a right triangle. Give the area for each.

1. base = 8 units	**2.** base = 10 units	**3.** base = 7 units
altitude = 4 units	altitude = 5 units	altitude = 6 units
area = _16 sq units_	area = _____	area = _____

Use graph paper to draw triangles, each with a different shape, having the given area. Give the base and altitude for each.

4. area = 15 sq units	**5.** area = 26 sq units	**6.** area = 18 sq units
base = _____	base = _____	base = _____
altitude = _____	altitude = _____	altitude = _____

Copyright © 1981 by Dale Seymour Publications.

66

Area of triangles

Name _____

Date _____

Find the base, altitude, and area of each triangle.

1. base = _13_
altitude = _4_
area = _26_

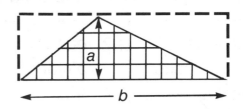

2. base = _____
altitude = _____
area = _____

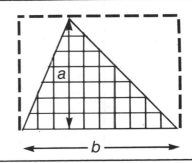

3. base = _____
altitude = _____
area = _____

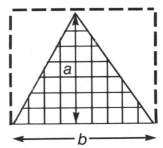

4. base = _____
altitude = _____
area = _____

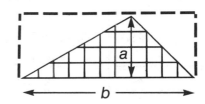

5. base = _____
altitude = _____
area = _____

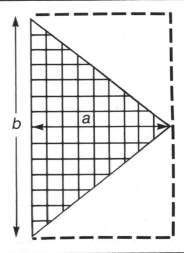

Copyright © 1981 by Dale Seymour Publications.

Drawing triangles

Name _____

Date _____

Use graph paper to draw triangles, each with a different shape, having the given base and altitude. Make one a right triangle. Give the area for each.

1. base = 9 units	2. base = 13 units	3. base = 7 units
altitude = 12 units	altitude = 8 units	altitude = 14 units
area = *54 sq units*	area = _____	area = _____

Use graph paper to draw triangles, each with a different shape, having the given area. Give the base and altitude for each.

4. area = 22 sq units	5. area = 14 sq units	6. area = 9 sq units
base = _____	base = _____	base = _____
altitude = _____	altitude = _____	altitude = _____

Copyright © 1981 by Dale Seymour Publications.

Area of triangles

Name _____

Date _____

Find the base, altitude, and area of each triangle.

1. base = _13_

altitude = _4_

area = _26_

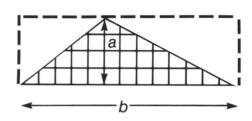

2. base = _____

altitude = _____

area = _____

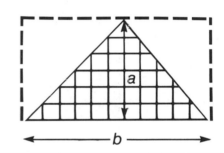

3. base = _____

altitude = _____

area = _____

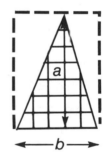

4. base = _____

altitude = _____

area = _____

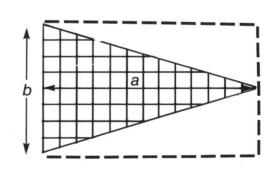

5. base = _____

altitude = _____

area = _____

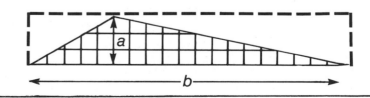

Copyright © 1981 by Dale Seymour Publications.

Drawing triangles

Name _____

Date _____

Use graph paper to draw triangles, each with a different shape, having the given base and altitude. Make one a right triangle. Give the area for each.

1. base = 9 units altitude = 4 units area = _18 sq units_	**2.** base = 12 units altitude = 5 units area = _____	**3.** base = 11 units altitude = 8 units area = _____

Use graph paper to draw triangles, each with a different shape, having the given area. Give the base and altitude for each.

4. area = 25 sq units base = _____ altitude = _____	**5.** area = 15 sq units base = _____ altitude = _____	**6.** area = 28 sq units base = _____ altitude = _____

Copyright © 1981 by Dale Seymour Publications.

Area of triangles
Area of parallelograms

Name _____

Date _____

Formulas for the area and perimeter of a triangle are as follows:

area of triangle $= \frac{1}{2}$ base × height   perimeter of triangle $=$ sum of sides

$A = \frac{1}{2} b \times h$ $P = a + b + c$

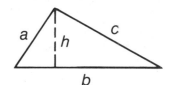

Find the area and perimeter of triangles given the following dimensions.

	a	b	c	h	area	perimeter
1.	40	39	25	24	468	104
2.	20	21	13	12		
3.	25	36	29	20		

Formulas for the area and perimeter of a parallelogram are as follows.

area of parallelogram $=$ base × height   perimeter of parallelogram $=$ sum of sides

$A = b \times h$ $P = 2a + 2b$

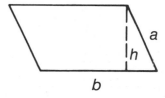

Find the area and perimeter of parallolograms given the following dimensions.

	a	b	h	area	perimeter
4.	15	20	9		
5.	18	31	12		
6.	30	27	18		

Copyright © 1981 by Dale Seymour Publications.

Area of trapezoids

The formulas for the area and perimeter of a trapezoid are as follows.

area of trapezoid= $\frac{1}{2}$ height × sum of bases

$$A = \tfrac{1}{2}h \times (b_1 + b_2)$$

perimeter of trapezoid= sum of sides

$$P = a + b_1 + c + b_2$$

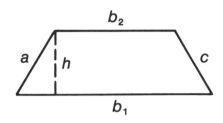

Find the area and perimeter of trapezoids given the following dimensions.

1. a = 22   area = __680__

   b_1 = 51   perimeter = __119__

   c = 29

   b_2 = 17 $A = \frac{1}{2}h \times (b_1 + b_2)$

   h = 20 $= \frac{1}{2} \times 20 \times (51 + 17)$

   $= 10 \times (68)$

   $= 680$

   $P = a + b_1 + c + b_2$

   $= 22 + 51 + 29 + 17$

   $= 119$

2. a = 15   area = _____

   b_1 = 35   perimeter = _____

   c = 13

   b_2 = 14

   h = 12

3. a = 85   area = _____

   b_1 = 145   perimeter = _____

   c = 39

   b_2 = 51

   h = 36

4. a = 53   area = _____

   b_1 = 96   perimeter = _____

   c = 35

   b_2 = 30

   h = 28

Copyright © 1981 by Dale Seymour Publications.

Name _____

Date _____

Formulas for the area and perimeter of a triangle are as follows:

area of triangle $= \frac{1}{2}$ base × height   perimeter of triangle $=$ sum of sides

$A = \frac{1}{2} b \times h$ $P = a + b + c$

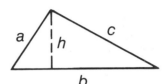

Find the area and perimeter of triangles given the following dimensions.

	a	b	c	h	area	perimeter
1.	100	91	61	60	2730	252
2.	123	156	45	27		
3.	13	40	37	12		

Formulas for the area and perimeter of a parallelogram are as follows.

area of parallelogram $=$ base × height   perimeter of parallelogram $=$ sum of sides

$A = b \times h$ $P = 2a + 2b$

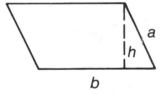

Find the area and perimeter of parallolograms given the following dimensions.

	a	b	h	area	perimeter
4.	20	35	16		
5.	17	27	14		
6.	34	32	19		

Copyright © 1981 by Dale Seymour Publications.   73

Area of trapezoids

The formulas for the area and perimeter of a trapezoid are as follows.

area of trapezoid $= \frac{1}{2}$ height $\times$ sum of bases

$$A = \frac{1}{2}h \times (b_1 + b_2)$$

perimeter of trapezoid $=$ sum of sides

$$P = a + b_1 + c + b_2$$

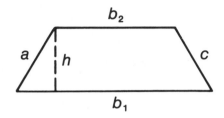

Find the area and perimeter of trapezoids given the following dimensions.

---

1. $a = 22$ area $= \underline{680}$
$b_1 = 51$ perimeter $= \underline{119}$
$c = 29$
$b_2 = 17$ $A = \frac{1}{2}h \times (b_1 + b_2)$
$h = 20$ $= \frac{1}{2} \times 20 \times (51 + 17)$
            $= 10 \times (68)$
            $= 680$

$P = a + b_1 + c + b_2$
    $= 22 + 51 + 29 + 17$
    $= 119$

---

2. $a = 41$ area $= \underline{\hspace{1cm}}$
$b_1 = 292$ perimeter $= \underline{\hspace{1cm}}$
$c = 181$
$b_2 = 72$
$h = 9$

---

3. $a = 89$ area $= \underline{\hspace{1cm}}$
$b_1 = 170$ perimeter $= \underline{\hspace{1cm}}$
$c = 65$
$b_2 = 38$
$h = 39$

---

4. $a = 104$ area $= \underline{\hspace{1cm}}$
$b_1 = 131$ perimeter $= \underline{\hspace{1cm}}$
$c = 41$
$b_2 = 26$
$h = 40$

Copyright © 1981 by Dale Seymour Publications.

Area of triangles
Area of parallelograms

Name _____

Date _____

Formulas for the area and perimeter of a triangle are as follows:

area of triangle $= \frac{1}{2}$ base $\times$ height   perimeter of triangle $=$ sum of sides

$A = \frac{1}{2} b \times h$ $P = a + b + c$

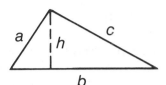

Find the area and perimeter of triangles given the following dimensions.

	a	b	c	h	area	perimeter
1.	25	28	17	15	210	70
2.	52	69	29	20		
3.	20	42	34	16		

Formulas for the area and perimeter of a parallelogram are as follows.

area of parallelogram $=$ base $\times$ height   perimeter of parallelogram $=$ sum of sides

$A = b \times h$ $P = 2a + 2b$

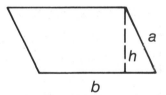

Find the area and perimeter of parallolograms given the following dimensions.

	a	b	h	area	perimeter
4.	15	16	7		
5.	24	28	16		
6.	26	31	21		

Copyright © 1981 by Dale Seymour Publications.

75

Area of trapezoids

Name _____

Date _____

The formulas for the area and perimeter of a trapezoid are as follows.

area of trapezoid= $\frac{1}{2}$ height $\times$ sum of bases

$$A = \frac{1}{2}h \times (b_1 + b_2)$$

perimeter of trapezoid= sum of sides

$$P = a + b_1 + c + b_2$$

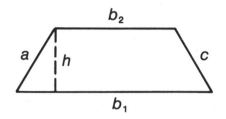

Find the area and perimeter of trapezoids given the following dimensions.

1. $a = 22$ area = __680__
$b_1 = 51$ perimeter = __119__
$c = 29$
$b_2 = 17$ $A = \frac{1}{2}h \times (b_1 + b_2)$
$h = 20$ $= \frac{1}{2} \times 20 \times (51 + 17)$
          $= 10 \times (68)$
          $= 680$

$P = a + b_1 + c + b_2$
   $= 22 + 51 + 29 + 17$
   $= 119$

2. $a = 20$ area = _____
$b_1 = 31$ perimeter = _____
$c = 37$
$b_2 = 20$
$h = 12$

3. $a = 109$ area = _____
$b_1 = 154$ perimeter = _____
$c = 68$
$b_2 = 31$
$h = 60$

4. $a = 65$ area = _____
$b_1 = 128$ perimeter = _____
$c = 20$
$b_2 = 53$
$h = 16$

Copyright © 1981 by Dale Seymour Publications.

Area of triangles
Area of parallelograms

Name _____

Date _____

Formulas for the area and perimeter of a triangle are as follows:

area of triangle = ½ base × height   perimeter of triangle = sum of sides

$A = \frac{1}{2}b \times h$ $P = a + b + c$

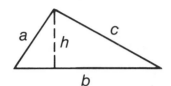

Find the area and perimeter of triangles given the following dimensions.

	a	b	c	h	area	perimeter
1.	20	51	37	12	306	108
2.	80	84	52	48		
3.	30	25	25	24		

Formulas for the area and perimeter of a parallelogram are as follows.

area of parallelogram = base × height   perimeter of parallelogram = sum of sides

$A = b \times h$ $P = 2a + 2b$

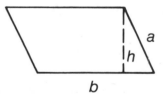

Find the area and perimeter of parallolograms given the following dimensions.

	a	b	h	area	perimeter
4.	18	16	15		
5.	21	29	18		
6.	28	31	24		

Copyright © 1981 by Dale Seymour Publications.

Area of trapezoids

The formulas for the area and perimeter of a trapezoid are as follows.

area of trapezoid = $\frac{1}{2}$ height × sum of bases

$$A = \frac{1}{2}h \times (b_1 + b_2)$$

perimeter of trapezoid = sum of sides

$$P = a + b_1 + c + b_2$$

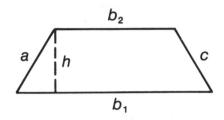

Find the area and perimeter of trapezoids given the following dimensions.

1. $a = 22$ area = **680**
 $b_1 = 51$ perimeter = **119**
 $c = 29$
 $b_2 = 17$ $A = \frac{1}{2}h \times (b_1 + b_2)$
 $h = 20$ $= \frac{1}{2} \times 20 \times (51 + 17)$
   $= 10 \times (68)$
   $= 680$
 	
 $P = a + b_1 + c + b_2$
  $= 22 + 51 + 29 + 17$
  $= 119$
	2. $a = 25$ area = _____
	 $b_1 = 51$ perimeter = _____
	 $c = 29$
	 $b_2 = 15$
	 $h = 20$

3. $a = 61$ area = _____ $b_1 = 54$ perimeter = _____ $c = 65$ $b_2 = 18$ $h = 60$	**4.** $a = 41$ area = _____ $b_1 = 276$ perimeter = _____ $c = 181$ $b_2 = 56$ $h = 9$

Copyright © 1981 by Dale Seymour Publications.

Area of triangles
Area of parallelograms

Name _____

Date _____

Formulas for the area and perimeter of a triangle are as follows:

$\begin{array}{l}\text{area of}\\ \text{triangle}\end{array} = \frac{1}{2}$ base × height $\begin{array}{l}\text{perimeter}\\ \text{of triangle}\end{array} =$ sum of sides

$A = \frac{1}{2}b \times h$ $P = a + b + c$

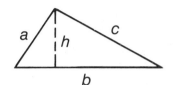

Find the area and perimeter of triangles given the following dimensions.

	a	b	c	h	area	perimeter
1.	68	53	61	60	1590	182
2.	75	92	29	21		
3.	15	44	37	12		

Formulas for the area and perimeter of a parallelogram are as follows.

$\begin{array}{l}\text{area of}\\ \text{parallelogram}\end{array} =$ base × height $\begin{array}{l}\text{perimeter of}\\ \text{parallelogram}\end{array} =$ sum of sides

$A = b \times h$ $P = 2a + 2b$

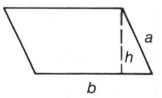

Find the area and perimeter of parallolograms given the following dimensions.

	a	b	h	area	perimeter
4.	21	17	19		
5.	25	35	17		
6.	36	42	31		

Copyright © 1981 by Dale Seymour Publications.   79

Area of trapezoids

The formulas for the area and perimeter of a trapezoid are as follows.

area of trapezoid= $\frac{1}{2}$ height × sum of bases

$$A = \tfrac{1}{2}h \times (b_1 + b_2)$$

perimeter of trapezoid= sum of sides

$$P = a + b_1 + c + b_2$$

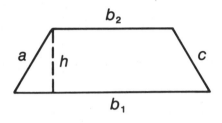

Find the area and perimeter of trapezoids given the following dimensions.

---

1. a = 22   area = __680__

  b_1 = 51   perimeter = __119__

  c = 29

  b_2 = 17 $A = \frac{1}{2}h \times (b_1 + b_2)$

  h = 20 $= \frac{1}{2} \times 20 \times (51 + 17)$

            $= 10 \times (68)$

            $= 680$

  $P = a + b_1 + c + b_2$

    $= 22 + 51 + 29 + 17$

    $= 119$

---

2. a = 99   area = _____

  b_1 = 187   perimeter = _____

  c = 48

  b_2 = 34

  h = 20

---

3. a = 58   area = _____

  b_1 = 239   perimeter = _____

  c = 150

  b_2 = 55

  h = 42

---

4. a = 58   area = _____

  b_1 = 77   perimeter = _____

  c = 41

  b_2 = 26

  h = 40

Copyright © 1981 by Dale Seymour Publications.

ANSWERS

Page 1   **1.** 10, 10, 40, 100 **2.** 10, 6, 32, 60 **3.** 13, 6, 38, 78 **4.** 13, 4, 34, 52 **5.** 6, 5, 22, 30

Page 2   **1.** 9, 3, 24, 27 **2.** 8, 4, 24, 32 **3.** 7, 2, 18, 14 **4.** 13, 6, 38, 78 **5.** 17, 2, 38, 34

Page 3   **1.** 12, 15, 54, 180 **2.** 3, 11, 28, 33 **3.** 9, 1, 20, 9 **4.** 3, 6, 18, 18 **5.** 14, 6, 40, 84

Page 4   **1.** 6, 11, 34, 66 **2.** 13, 11, 48, 143 **3.** 11, 5, 32, 55 **4.** 8, 7, 30, 56 **5.** 22, 5, 54, 110

Page 5   **1.** 12, 6, 36, 72 **2.** 5, 10, 30, 50 **3.** 27, 3, 60, 81 **4.** 5, 11, 32, 55 **5.** 4, 9, 26, 36

Page 6   **1.** 6, 15, 42, 90 **2.** 10, 17, 54, 170 **3.** 9, 12, 42, 108 **4.** 2, 11, 26, 22 **5.** 12, 5, 34, 60

Page 7   **1.** 11, 16, 54, 176 **2.** 7, 11, 36, 77 **3.** 8, 5, 26, 40 **4.** 8, 6, 28, 48 **5.** 17, 2, 38, 34

Page 8   **1.** 7, 14, 42, 98 **2.** 1, 8, 18, 8 **3.** 5, 6, 22, 30 **4.** 17, 5, 44, 85 **5.** 25, 6, 62, 150

Page 9   **1.** 11, 10, 42, 110 **2.** 10, 8, 36, 80 **3.** 14, 16, 60, 224 **4.** 6, 12, 36, 72 **5.** 21, 2, 46, 42

Page 10   **1.** 12, 12, 48, 144 **2.** 14, 1, 30, 14 **3.** 10, 4, 28, 40 **4.** 7, 6, 26, 42 **5.** 16, 4, 40, 64

Page 11   **1.** 6, 2, 16; 4, 3, 14; 12, 1, 26 **2.** 20, 1, 42; 10, 2, 24; 5, 4, 18 **3.** 24, 1, 50; 12, 2, 28; 8, 3, 22; 4, 6, 20 **4.** 18, 1, 38; 9, 2, 22; 6, 3, 18 **5.** 30, 1, 62; 15, 2, 34; 10, 3, 26; 6, 5, 22 Examples are given for problems **6-10.** **6.** 8, 7, 56; 9, 6, 54; 10, 5, 50 **7.** 18, 1, 18; 17, 2, 34; 16, 3, 48 **8.** 13, 1, 13; 12, 2, 24; 11, 3, 33 **9.** 10, 1, 10; 9, 2, 18; 8, 3, 24 **10.** 12, 1, 12; 11, 2, 22; 10, 3, 30

Page 12   **1.** 7, 4 **2.** 9, 4 **3.** 8, 5 **4.** 8, 7 **5.** 9, 5

Page 13   **1.** 10, 6, 32; 12, 5, 34; 15, 4, 38; 20, 3, 46; 30, 2, 64; 60, 1, 122 **2.** 10, 10, 40; 50, 2, 104; 100, 1, 202; 25, 4, 58; 20, 5, 50 **3.** 8, 4, 24; 16, 2, 36; 32, 1, 66 **4.** 7, 6, 26; 21, 2, 46; 14, 3, 34; 42, 1, 86 **5.** 10, 5, 30; 25, 2, 54; 50, 1, 102 Examples are given for problems **6-10.** **6.** 19, 1, 19; 18, 2, 36; 17, 3, 51 **7.** 27, 1, 27; 26, 2, 52; 25, 3, 75 **8.** 6, 1, 6; 5, 2, 10; 4, 3, 12 **9.** 15, 1, 15; 14, 2, 28; 13, 3, 39 **10.** 22, 1, 22; 21, 2, 42; 20, 3, 60

Page 14   **1.** 12, 3 **2.** 12, 2 **3.** 15, 3 **4.** 6, 5 **5.** 5, 4

Page 15   **1.** 28, 1, 58; 14, 2, 32; 7, 4, 22 **2.** 32, 1, 66; 16, 2, 36; 8, 4, 24 **3.** 30, 1, 62; 15, 2, 34; 10, 3, 26; 6, 5, 22 **4.** 42, 1, 86; 21, 2, 46; 14, 3, 34; 7, 6, 26 **5.** 75, 1, 152; 25, 3, 56; 15, 5, 40 Examples are given for problems **6-10.** **6.** 5, 4, 20; 8, 1, 8; 7, 2, 14 **7.** 5, 5, 25; 6, 4, 24; 7, 3, 21 **8.** 16, 1, 16; 15, 2, 30; 14, 3, 42 **9.** 23, 1, 23; 22, 2, 44; 21, 3, 63 **10.** 10, 1, 10; 9, 2, 18; 8, 3, 24

Page 16   **1.** 5, 5 **2.** 18, 5 **3.** 11, 4 **4.** 8, 3 **5.** 12, 5

Page 17   **1.** 36, 1, 74; 18, 2, 40; 12, 3, 30; 9, 4, 26; 6, 6, 24 **2.** 30, 1, 62; 15, 2, 34; 10, 3, 26; 6, 5, 22 **3.** 40, 1, 82; 8, 5, 26; 20, 2, 44; 10, 4, 28 **4.** 72, 1, 146; 36, 2, 76; 18, 4, 44; 12, 6, 36; 9, 8, 34; 24, 3, 54 **5.** 54, 1, 110; 27, 2, 58; 18, 3, 42; 9, 6, 30 Examples are given for problems **6-10.** **6.** 7, 1, 7; 6, 2, 12; 5, 3, 15 **7.** 11, 1, 11; 10, 2, 20; 9, 3, 27 **8.** 17, 1, 17; 16, 2, 32; 15, 3, 45 **9.** 13, 1, 13; 12, 2, 24; 11, 3, 33 **10.** 15, 1, 15; 14, 2, 28; 13, 3, 39

Page 18   **1.** 20, 3 **2.** 9, 9 **3.** 6, 6 **4.** 20, 4 **5.** 12, 7

Page 19   **1.** 20, 1, 42; 10, 2, 24; 5, 4, 18 **2.** 45, 1, 92; 15, 3, 36; 9, 5, 28 **3.** 32, 1, 66; 16, 2, 36; 8, 4, 24 **4.** only two rectangles possible 27, 1, 56; 9, 3, 24 **5.** 18, 1, 38; 9, 2, 22; 6, 3, 18 Examples are given for problems **6-10.** **6.** 5, 1, 5; 4, 2, 8; 3, 3, 9 **7.** 8, 1, 8; 7, 2, 14; 6, 3, 18 **8.** 10, 1, 10; 9, 2, 18; 8, 3, 24 **9.** 7, 1, 7; 6, 2, 12; 5, 3, 15 **10.** 6, 1, 6; 5, 2, 10; 4, 3, 12

Page 20   **1.** 8, 5 **2.** 6, 4 **3.** 8, 8 **4.** 8, 7 **5.** 8, 2

Page 21   **1.** 72 ft, 288 ft² **2.** 90 yd, 350 yd² **3.** 168 in., 1715 in.² **4.** 70 yd, 216 yd² **5.** 41 ft, 102 ft²

Page 22   **1.** 4.32 m, 0.494 m² **2.** 4.04 cm, 0.544 cm² **3.** 64.046 m, 0.736m² **4.** 5.82 km, 1.025 km² **5.** 3.74 cm, 0.432 cm²

Page 23   **1.** 106 in., 630 in.² **2.** 78 ft, 374 ft² **3.** 32 ft, 64 ft² **4.** $34\frac{2}{3}$ ft, 75 ft² **5.** $26\frac{1}{2}$ ft, 42 ft²

81

Page 24   **1.** 33 dm, 66.96 dm²   **2.** 358 cm, 8004 cm²   **3.** 9.6 m, 3.84 m²   **4.** 249 m, 305 m²   **5.** 25.3 m, 14.25 m²

Page 25   **1.** 40 ft, 96 ft²   **2.** 48 ft, 135 ft²   **3.** $42\frac{1}{2}$ ft, 111 ft²   **4.** $26\frac{4}{9}$ yd, $43\frac{1}{3}$ yd²   **5.** 82 in., 408 in.²

Page 26   **1.** 14 dm, 9.01 dm²   **2.** 11 m, 5.16 m²   **3.** 174.4 mm, 103.2 mm²   **4.** 346 m, 3450 m²   **5.** 28.000046 m, 0.000322 m²

Page 27   **1.** 62 in., 238 in.²   **2.** 35 ft, 66 ft²   **3.** 154 ft, 1470 ft²   **4.** 56 ft, 195 ft²   **5.** $66\frac{1}{3}$ ft, 273 ft²

Page 28   **1.** 35.8 dm, 80.04 dm²   **2.** 14.2 m, 12.04 m²   **3.** 22.8 cm, 24.65 cm²   **4.** 16.66 m, 1.863 m²   **5.** 1346 cm, 14,950 cm²

Page 29   **1.** 136 in., 1075 in.²   **2.** 338 ft, 7008 ft²   **3.** 56 yd, 171 yd²   **4.** 69 ft, 182 ft²   **5.** $42\frac{1}{2}$ ft, 84 ft²

Page 30   **1.** 3.76 cm, 0.448 cm²   **2.** 29 mm, 50.74 mm²   **3.** 2890 cm, 165,000 cm²   **4.** 256 cm, 4095 cm²   **5.** 164.0086 m, 0.3526 m²

Page 31   **1.** 76 ft, 240 ft²   **2.** 88 yd, 384 yd²   **3.** 176 in., 1152 in.²   **4.** 82 in., 378 in.²   **5.** 154 ft, 930 ft²   **6.** $21\frac{3}{10}$ mi, $27\frac{1}{5}$ mi²   **7.** $21\frac{1}{6}$ ft, 28 ft²   **8.** 26 in., 90 in.   **9.** 18 ft, 94 ft   **10.** 105 ft, 1470 ft²   **11.** 128 yd, 4608 yd²

Page 32   **1.** 43 cm, 113.16 cm²   **2.** 22.66 m, 29.3532 m²   **3.** 5.7 m, 17.8 m   **4.** 4.03 mm, 146.06 mm   **5.** 2.6 cm, 9.1 cm²   **6.** 9.3 m, 80.91 m²   **7.** 4.14 cm, 100.28 cm

Page 33   **1.** 40 ft, 75 ft²   **2.** 56 in., 192 in.²   **3.** 120 yd, 800 yd²   **4.** 76 ft, 345 ft²   **5.** 98 in., 544 in.²   **6.** $11\frac{1}{6}$ ft, $7\frac{1}{2}$ ft²   **7.** $27\frac{2}{3}$ yd, $47\frac{1}{2}$ yd²   **8.** 7 ft, 52 ft   **9.** 12 in., 76 in.   **10.** 35 yd, 840 yd²   **11.** 32 ft, 1824 ft²

Page 34   **1.** 310 cm, 5976 cm²   **2.** 186 mm, 2052 mm²   **3.** 37 m, 84.46 m²   **4.** 18.2 dm, 19.38 dm²   **5.** 2.6 m, 14.8 m   **6.** 3.59 cm, 17.4 cm   **7.** 5.4 m, 21.06 m²

Page 35   **1.** 32 yd, 48 yd²   **2.** 120 ft, 675 ft²   **3.** 118 ft, 864 ft²   **4.** 178 yd, 1908 yd²   **5.** $26\frac{1}{6}$ ft, 35 ft²   **6.** $24\frac{5}{6}$ ft, $37\frac{5}{8}$ ft²   **7.** $29\frac{1}{6}$ yd, 49 yd²   **8.** 18 in., 82 in.   **9.** 27 in., 126 in.   **10.** 31 ft, 1457 ft²   **11.** $5\frac{1}{2}$ ft, $30\frac{1}{4}$ ft²

Page 36   **1.** 100 cm, 544 cm²   **2.** 220 cm, 2881 cm²   **3.** 230 m, 2116 m²   **4.** 12.24 mm, 6.336 mm²   **5.** 61 cm, 306 cm   **6.** 4.3 m, 19.8 m   **7.** 36 mm, 1548 mm²

Page 37   **1.** 82 ft, 364 ft²   **2.** 148 ft, 928 ft²   **3.** 212 in., 2448 in.²   **4.** 166 yd, 1482 yd²   **5.** $35\frac{2}{3}$ ft, 77 ft²   **6.** $22\frac{5}{6}$ ft, $31\frac{2}{3}$ ft²   **7.** $41\frac{1}{6}$ yd, $104\frac{5}{6}$ yd²   **8.** 28 in., 142 in.   **9.** 43 in., 220 in.   **10.** 27 in., 378 in.²   **11.** 5 ft, 75 ft²

Page 38   **1.** 40 m, 75 m²   **2.** 72 cm, 288 cm²   **3.** 79.8 cm, 107.3 cm²   **4.** 73 cm, 248 cm   **5.** 96 mm, 318 mm   **6.** 34 cm, 2618 cm²   **7.** 26 mm, 988 mm²

Page 39   **1.** 120 ft, 576 ft²   **2.** 112 in., 703 in.²   **3.** 136 in., 1131 in.²   **4.** 216 ft, 2432 ft²   **5.** 278 yd, 1750 yd²   **6.** $15\frac{1}{6}$ ft, 12 ft²   **7.** $15\frac{1}{6}$ ft, $14\frac{1}{6}$ ft²   **8.** 22 yd, $25\frac{5}{9}$ yd²   **9.** 13 in., 84 in.   **10.** 24 ft, 220 ft   **11.** 17 in., 100 in.

Page 40   **1.** 168 mm, 1508 mm²   **2.** 51.4 cm, 136.5 cm²   **3.** 4.454 m, 1.2051 m²   **4.** 27.84 mm, 45.752 mm²   **5.** 34.5 cm, 253 cm   **6.** 203 m, 14,007 m²   **7.** 27 mm, 1674 mm²

Page 41   **1.** 11, 4, 22   **2.** 12, 5, 30   **3.** 10, 7, 35   **4.** 8, 10, 40   **5.** 7, 8, 28   **6.** 5, 6, 15   **7.** 8, 12, 48   **8.** 14, 8, 56

Page 42   **1.** 16, 8, 64   **2.** 11, 6, 33   Answers for **3-8** are examples.   **3.** 9, 8   **4.** 8, 9   **5.** 12, 6   **6.** 4, 12   **7.** 12, 4   **8.** 16, 3

Page 43   **1.** 11, 6, 33   **2.** 7, 4, 14   **3.** 10, 8, 40   **4.** 6, 8, 24   **5.** 17, 6, 51   **6.** 5, 6, 15   **7.** 14, 5, 35   **8.** 3, 5, $7\frac{1}{2}$

Page 44   **1.** 5, 4, 10   **2.** 6, 8, 24   Answers for **3-8** are examples.   **3.** 6, 10   **4.** 10, 6   **5.** 5, 12   **6.** 8, 10   **7.** 10, 8   **8.** 16, 5

Page 45   **1.** 13, 6, 39   **2.** 3, 8, 12   **3.** 10, 4, 20   **4.** 8, 8, 32   **5.** 6, 7, 21   **6.** 16, 3, 24   **7.** 12, 4, 24   **8.** 8, 6, 24

Page 46   **1.** 9, 7, $31\frac{1}{2}$ **2.** 15, 8, 60   Answers for **3-8** are examples.   **3.** 4, 6   **4.** 6, 4   **5.** 3, 8   **6.** 8, 2   **7.** 2, 8   **8.** 4, 4

Page 47   **1.** 13, 6, 39   **2.** 4, 6, 12   **3.** 9, 6, 27   **4.** 8, 6, 24   **5.** 11, 9, $49\frac{1}{2}$ **6.** 11, 4, 22   **7.** 8, 7, 28   **8.** 11, 4, 22

82

Page 48   **1.** 4, 4, 8   **2.** 8, 9, 36   Answers for 3-8 are examples.   **3.** 4, 5   **4.** 5, 4   **5.** 2, 10   **6.** 4, 10   **7.** 10, 4   **8.** 8, 5

Page 49   **1.** 11, 9, $49\frac{1}{2}$ **2.** 12, 3, 18   **3.** 9, 7, $31\frac{1}{2}$ **4.** 14, 10, 70   **5.** 9, 4, 18   **6.** 7, 9, $31\frac{1}{2}$ **7.** 15, 3, $22\frac{1}{2}$ **8.** 9, 4, 18

Page 50   **1.** 14, 2, 14   **2.** 10, 5, 25   Answers for 3-8 are examples.   **3.** 4, 9   **4.** 9, 4   **5.** 6, 6   **6.** 4, 8   **7.** 8, 4   **8.** 16, 2

Page 51   **1.** 210 in.², 70 in.   **2.** 630 yd², 126 yd   **3.** 1320 ft², 176 ft   **4.** 2730 in.², 260 in.   **5.** 840 yd², 140 yd

Page 52   **1.** 924 km², 154 km   **2.** 330 m², 132 m   **3.** 5670 cm², 378 cm   **4.** 1560 cm², 208 cm   **5.** 0.0546 m², 1.82 m

Page 53   **1.** 990 ft², 220 ft   **2.** 840 in.², 140 in.   **3.** 150 yd², 60 yd   **4.** 1386 in.², 198 in.   **5.** 924 ft², 154 ft

Page 54   **1.** 7.26 cm², 13.2 cm   **2.** 19.2 m², 24 m   **3.** 29.4 dm², 28 dm   **4.** 68.04 m², 50.4 m   **5.** 6480 cm², 540 cm

Page 55   **1.** 120 ft², 60 ft   **2.** 486 ft², 108 ft   **3.** 4320 in.², 360 in.   **4.** $52\frac{1}{2}$ yd², 35 yd   **5.** 1620 yd², 270 yd

Page 56   **1.** 7.26 cm², 13.2 cm   **2.** 48.6 mm², 36 mm   **3.** 0.5376 km², 4.48 km   **4.** 72.6 dkm², 44 dkm   **5.** 0.135 km², 1.8 km

Page 57   **1.** 216 in.², 72 in.   **2.** 384 ft², 96 ft   **3.** 11,520 yd², 720 yd   **4.** 1920 in.², 240 in.   **5.** 336 ft², 112 ft

Page 58   **1.** 19.44 m², 21.6 m   **2.** 0.588 km² 4.2 km   **3.** 8.64 km², 14.4 km   **4.** 75.6 dm², 42 dm   **5.** 10.14 m², 15.6 m

Page 59   **1.** 240 ft², 80 ft   **2.** 720 in.², 180 in.   **3.** 1386 yd², 198 yd   **4.** 2340 ft², 234 ft   **5.** 546 in.², 182 in.

Page 60   **1.** 294 cm², 84 cm   **2.** 336 mm², 112 mm   **3.** 9.24 cm², 15.4 cm   **4.** 0.033 dm², 1.32 dm   **5.** 156,000 cm², 2080 cm

Page 61   **1.** 9, 4, 18   **2.** 12, 5, 30   **3.** 10, 7, 35   **4.** 9, 3, $13\frac{1}{2}$ **5.** 10, 5, 25

Page 62   Examples are given.   **1.** 15 sq units   **2.** 18 sq units   **3.** 18 sq units   **4.** 4, 7   **5.** 4, 8   **6.** 5, 4

Page 63   **1.** 11, 5, $27\frac{1}{2}$ **2.** 12, 7, 42   **3.** 11, 6, 33   **4.** 8, 5, 20   **5.** 14, 3, 21

Page 64   Examples are given.   **1.** 14 sq units   **2.** 45 sq units   **3.** 72 sq units   **4.** 8, 5   **5.** 9, 4   **6.** 3, 8

Page 65   **1.** 13, 5, $32\frac{1}{2}$ **2.** 7, 7, $24\frac{1}{2}$ **3.** 5, 6, 15   **4.** 14, 8, 56   **5.** 17, 7, $59\frac{1}{2}$

Page 66   Examples are given.   **1.** 16 sq units   **2.** 25 sq units   **3.** 21 sq units   **4.** 6, 5   **5.** 4, 13   **6.** 6, 6

Page 67   **1.** 13, 4, 26   **2.** 10, 7, 35   **3.** 9, 7, $31\frac{1}{2}$ **4.** 11, 4, 22   **5.** 14, 9, 63

Page 68   Examples are given.   **1.** 54 sq units   **2.** 52 sq units   **3.** 49 sq units   **4.** 4, 11   **5.** 4, 7   **6.** 3, 6

Page 69   **1.** 13, 4, 26   **2.** 11, 6, 33   **3.** 5, 7, $17\frac{1}{2}$ **4.** 8, 13, 52   **5.** 19, 3, $28\frac{1}{2}$

Page 70   Examples are given.   **1.** 18 sq units   **2.** 30 sq units   **3.** 44 sq units   **4.** 10, 5   **5.** 6, 5   **6.** 7, 8

Page 71   **1.** 468, 104   **2.** 126, 54   **3.** 360, 90   **4.** 180, 70   **5.** 372, 98   **6.** 486, 114

Page 72   **1.** 680, 119   **2.** 294, 77   **3.** 3528, 320   **4.** 1764, 214

Page 73   **1.** 2730, 252   **2.** 2106, 324   **3.** 240, 90   **4.** 560, 110   **5.** 378, 88   **6.** 608, 132

Page 74   **1.** 680, 119   **2.** 1638, 586   **3.** 4056, 362   **4.** 3140, 302

Page 75   **1.** 210, 70   **2.** 690, 150   **3.** 336, 96   **4.** 112, 62   **5.** 448, 104   **6.** 651, 114

Page 76   **1.** 680, 119   **2.** 306, 108   **3.** 5550, 362   **4.** 1448, 266

Page 77   **1.** 306, 108   **2.** 2016, 216   **3.** 300, 80   **4.** 240, 68   **5.** 522, 100   **6.** 744, 118

Page 78   **1.** 680, 119   **2.** 660, 120   **3.** 2160, 198   **4.** 1494, 554

Page 79   **1.** 1590, 182   **2.** 966, 196   **3.** 264, 96   **4.** 323, 76   **5.** 595, 120   **6.** 1302, 156

Page 80   **1.** 680, 119   **2.** 2210, 368   **3.** 6174, 502   **4.** 2060, 202